Ultimat
Operat
Handbook
Robinson R44

by Bastian Liebermann

bajali Publishing LLC
bajalipublishing.com

Ultimate Pilot's Operating Handbook - Robinson R44
Bastian Liebermann

© Copyright 2011, Bastian Liebermann

Published by
bajali Publishing LLC

info@bajalipublishing.com
www.bajalipublishing.com

ISBN 978-0-9836962-0-9

Printed in the United States of America
Hillsboro, OR

About the Author

Bastian Liebermann was born in the City of New York (USA). After he spent some of his childhood years there he and his family moved back to Germany, where he then lived for the remaining part of his youth and young adulthood.

After graduating from school he then worked as a media designer for about 2 years before he pursued his dream of flying helicopters. He began his aviation career by getting his private pilot's license for helicopters in Germany.

Having done some flying in Germany he then decided to go to the US to continue his training. Now he holds the FAA commercial pilot's license with an instrument rating and also is an instrument-rated flight instructor.

He still keeps his German license valid as well and enjoys being able to fly both in the US and in Europe when he visits family and friends there.

For more information about flight training, flight safety and general aviation be sure to visit my website CFIBastian.com

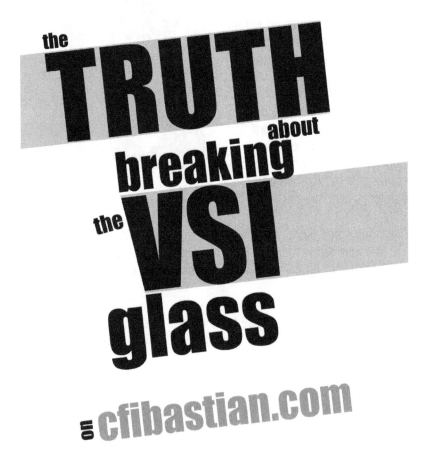

Thanks to:

Daniel Huesca,
Dirk Herr,
Hieronymus Sarholz,
Hillsboro Aviation,
The Federal Aviation Administration of the USA,
Jan Veen,
N.Mark Sylvester from ASAP Avionics,
Mark Hohstadt,
Markus Grutke,
Michael Meier,
Robinson Helicopter Company,
Taylor Rohde,
Tim Tucker,
Tim McAdams,
Torsten Wehner,
and my family and friends

for the continuously great support regarding this project.

Table of Contents

General

Limitations

Emergency Procedures

Normal Procedures

Checklists

Loading & Performance

Systems Description

Awareness Training

Carb-Icing Supplement

Conversions

Chapter 1

General

Chapter 1
General

About the Robinson 44

With the Robinson R22 having revolutionized the market for light training helicopters, the Robinson Helicopter Company decided to take it a step further and develop their second helicopter model - the R44.

As with the R22 the R44 soon began to revolutionize another niche of the helicopter market, which had no or very little competition to the Robinson models.
While the R22 aimed at flight schools to provide a light and inexpensive training helicopter, the R44 kept to those advantages and aimed at touring companies and light utility companies which so far went with bigger, more expensive and demanding helicopters.

With the Astro the first version in the series of the R44 models was introduced in 1992. Over the next decade other versions followed its footsteps beginning with the introduction of the R44 Clipper Version, Float-equipped helicopters for amphibious usage in 1996.

In 1997 and 1998 Robinson introduced two special versions of the R44 line, which are the R44 Police Helicopter, especially equipped for the needs of law enforcement agencies, and the R44 Newscopter that is designed to provide an airborne, broadcast-quality studio.

With the turn of the Millennium a new main model was introduced, superseding the Astro model. The Raven I features a standard hydraulic control system, which greatly improved the flying experience for the pilot compared to the electric trim motors that were used in the Astro models. It also comes with adjustable pedals on the pilot's side.

Two years later, in 2002, Robinson introduced an upgraded version of the Raven model, the Raven II, which is sold alongside the Raven I.
The Raven II is fuel-injected, has more power and a higher gross weight than the carbureted Raven I.

Although specifically tailored to better handle High-Altitude operations many operators prefer the Raven II for usage because of the added power and, given the fuel-injected engine, there is no need to worry about carb-ice anymore.
Both the Raven I and II are available as float-equipped versions - then called Clipper I and II.

The Raven II is also available as an IFR Trainer, which is equipped with the necessary instruments for IFR helicopter training.

In October 2002, the R44 became the first piston helicopter to fly to the North Pole, piloted by Quentin Smith and Steve Brooks. Three years later, the same two pilots flew an R44 Raven II to the South Pole - another first piston helicopter record for the R44.

3-Side View of the R44

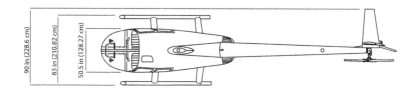

198 in (502.92 cm)

58 in (147.32 cm)

72 in (182.88 cm)

30 in (76.2 cm) MINIMUM TAIL
SKID HEIGHT MEASURED FROM
LEVEL GROUND UP

353 in (896.62 cm)

459 in (1165.86 cm)

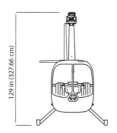

129 in (327.66 cm)

3-Side View of the R44 Clipper

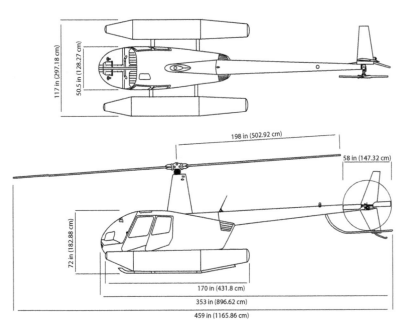

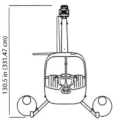

Timeline and Models
of the Robinson 44

General

R44 Astro	1992 - 2000
R44 Raven I	2000 - present
R44 Raven II	2002 - present
R44 IFR Trainer	1995 - present
R44 Clipper	1996 - 2000
R44 Clipper I	2000 - present
R44 Clipper II	2002 - present
R44 Police Helicopter	1997 - present
R44 Newscopter	1998 - present

As this book is published in 2011, the term "present" refers to the year 2011.

R44 Astro

R44 Raven I

R44 Raven II

R44 Clipper with pop out floats

R44 Clipper with fixed floats

R44 Newscopter

R44 Police Helicopter

Chapter 2

Limitations

Chapter 2
Limitations

Color Code for Instrument Markings

Red　　　　　Indicates operating limits. Pointer should not move past the red line during normal operation

Red Cross-hatch　　　　　Indicates power-off V_{NE}

Yellow　　　　　Precautionary or special operating procedure range

Green　　　　　Normal operating range

In the R44 the speed limitations for operation up to 3000 feet density altitude are marked on the Airspeed Indicator as follows:

Green Arc (Normal operation range):　　　　0 to 130 KIAS

Red Line (Normal operation V_{NE}):　　　　130 KIAS

Red cross-hatch Line (Power-off V_{NE}):　　　　100 KIAS

V_{NE} : Never Exceed Speed

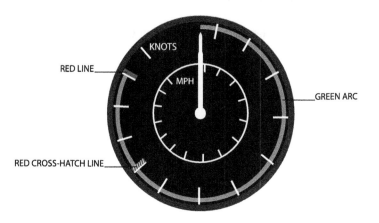

Airspeed Limits

Never exceed speed (V_{NE}):

Up to 3000 feet density altitude there are set never exceed speeds, whereas they vary above this altitude.

Airspeed Limits UP TO 3000 FT DA	R44 - No floats	R44 - With floats
2200 lbs. TOGW & below	130 KIAS	120 KIAS
Over 2200 lbs. TOGW	120 KIAS	110 KIAS
Autorotation	100 KIAS	100 KIAS

The Abbreviation "TOGW" stands for Take Off Gross Weight.

When flying above 3000 feet refer to the never exceed speed table to calculate the appropriate V_{NE} speed.

Do not exceed 100 KIAS when operating with power higher than Maximum Continuous Power (MCP).

Do not exceed 100 KIAS when operating with any door(s) removed.

Rotorspeed Limits

	Tachometer reading	Actual RPM
Power On		
Maximum	102 %	408
Minimum	101 %	404
Minimum (Early models)	99 %	396

	Tachometer reading	Actual RPM
Power Off		
Maximum	108 %	432
Minimum	90 %	360

Tachometer Instrument Markings

	Indication	Denotation
Rotor Tachometer		
Upper red line	108 %	Maximum RPM
Green Arc	90 to 108 %	Normal Operating Range
Lower red line	90 %	Minimum RPM

	Indication	Denotation
Engine Tachometer		
Upper red line	102 %	Max Engine Speed
Green Arc	101 to 102 % 99 to 102 % (old models)	Normal Operating Range
Lower red line	101 % 99 % (old models)	Minimum Engine Speed

Engine Instrument Markings

	Indication	Denotation
Oil Pressure		
Lower red line	25 psi	Minimum during idle
Lower yellow arc	25 to 55 psi	Idle
Green arc	55 to 95 psi	Normal Operating Range
Upper yellow arc	95 to 115 psi	Start and Warm-Up
Upper red line	115 psi	Maximum Start and Warm-up
Oil Temperature		
Green arc	75 to 245°F (24 to 118°C)	Normal Operating Range
Red line	245°F (118°C)	Maximum Temperature
Cylinder Head Temperature		
Green arc	200 to 500°F (93 to 260°C)	Normal Operating Range
Red line	500°F (260°C)	Maximum Temperature
Manifold Pressure	**Raven II**	**Astro and Raven I**
Green arc	15.0 to 23.3 in. HG	16.0 to 24.7 in. HG
Yellow arc	19.1 to 26.1 in. HG	21.8 to 26.3 in. HG
Red line	26.1 in. HG	26.3 in. HG

The yellow manifold pressure arc denotes variable MAP limits.
Refer to the Limit Manifold Pressure table.

Weight Limits

	Lbs.	KG
Raven II		
Max Gross Weight	2500	1134
Min Gross Weight	1600	726
Maximum per seat including baggage compartment	300	136
Maximum in any baggage compartment	50	23
Astro + Raven I		
Max Gross Weight	2400	1089
Min Gross Weight	1550	703
Maximum per seat including baggage compartment	300	136
Maximum in any baggage compartment	50	23

For all models:
Minimum solo pilot plus forward baggage weight with all doors installed is 150 lbs. (68 kg) unless a weight and balance computation shows CG is within limits.
Ballast may be required.

For more detailed Information about loading of the Newscopter or the Police Helicopter version of the R44 see the "Loading & Performance" Chapter.

Center of Gravity Limits

The datum line is 100 inches forward of the main rotor shaft centerline.

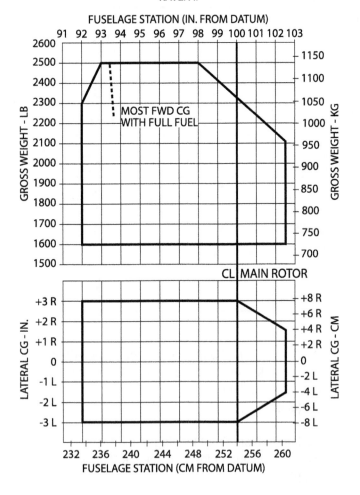

CENTER OF GRAVITY LIMITS
RAVEN II

Note: Weight and Balance Limitations for Clipper models are is-sued for each specific helicopter only. Refer to the supplied POH, if flying a Clipper.

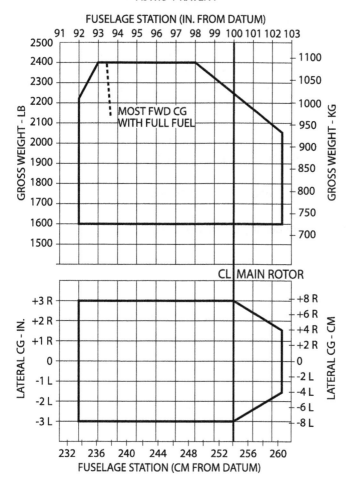

CENTER OF GRAVITY LIMITS
ASTRO + RAVEN I

Flight and Maneuver Limitations

Aerobatic flight is prohibited.

Low-G cyclic pushovers are prohibited.

Caution:
A pushover (forward cyclic maneuver) performed from level flight or following a pull-up causes a low-G (near weightless) condition, which can result in catastrophic loss of lateral control. To eliminate a low-G condition, immediately apply gentle aft cyclic. Should a right roll commence during a low-G condition, apply gentle aft cyclic to reload the rotor before applying lateral cyclic to stop the roll.

Flight is prohibited with governor selected off, with exceptions for in-flight system malfunction or emergency procedures training.

Flight in known icing conditions is prohibited.

Maximum operating density altitude is 14,000 feet.

Maximum operating altitude is 9000 feet AGL to allow for landing within 5 minutes in case of fire.

Alternator, RPM governor, low rotor RPM warning system, OAT gage and hydraulic control system must be operational for dispatch.

Solo flight is allowed from the right seat only.

Forward left seat belt must be buckled.

Minimum crew is one pilot.

Doors-off operation up to 100 KIAS is approved with any or all doors removed.

Caution:
No loose items allowed in cabin during doors-off flight.

Caution:
Avoid abrupt control inputs. They produce high fatigue stress and could lead to a premature and catastrophic failure of a critical component.

Operation Limitations

VFR day is approved.

VFR operation at night is permitted only when landing, navigation, instrument and anti-collision lights are operational. Orientation during night flight must be maintained by visual reference to ground objects illuminated by lights on the ground or adequate celestial illumination.

Note:
In countries outside the U.S. there may
be different operating limitations.

Fuel Capacity

	Total Capacity US Gallons (liters)	Useable Capacity US Gallons (liters)
Tank with bladders		
Main	30.5 (115)	29.5 (112)
Aux	17.2 (65)	17.0 (64)
Total	47.7 (180)	46.5 (176)
Tank without bladders		
Main	31.6 (120)	30.6 (116)
Aux	18.5 (70)	18.3 (69)
Total	50.1 (190)	48.9 (185)

Approved fuel grades are:
100 LL GRADE AVIATION FUEL
100/130 GRADE AVIATION FUEL

Oil Quantity

Oil Quantity	US Quarts	Liters
Capacity	12	11.4
Minimum for takeoff	7	6.6

Recommended Oil Grades

Average Ambient Air Temperature	Mineral Grades MIL-L-6082 or SAE J 1966 (Use first 50 hours)	Ashless Dispersant Grades MIL-L-22851 or SAE J 1899 (Use after first 50 hours)
All Temperature	--	SAE 15W-50 or SAE 20W-50
Above 80°F (27°C)	SAE 60	SAE 60
Above 60°F (16°C)	SAE 50	SAE 40 or SAE 50
30°F to 90°F (-1°C to 32°C)	SAE 40	SAE 40
0°F to 70°F (18°C to 21°C)	SAE 30	SAE 30, SAE 40 or SAE 20W-40
0°F to 90°F (18°C to 32°C)	SAE 20W-50	SAE 20W-50 or SAE 15W-50
Below 10°F (12°C)	SAE 20	SAE 30 or SAE 20W-30

Never Exceed Speed and Manifold Pressure Limitations Tables

RAVEN II
LIMIT MANIFOLD PRESSURE - IN. HG

PRESS	MAXIMUM CONTINOUS POWER							
	OAT - °C							
ALT-FT	-30	-20	-10	0	10	20	30	40
SL	21.5	21.8	22.1	22.4	22.6	22.9	23.1	23.3
2000	20.9	21.2	21.5	21.8	22.1	22.3	22.5	22.8
4000	20.4	20.7	21.0	21.3	21.5	21.8	22.0	22.2
6000	19.9	20.2	20.5	20.8	21.0	21.3	21.5	21.7
8000	19.5	19.8	20.1	20.3	20.6	20.8	21.0	21.3
10000	19.1	19.4	19.6	19.9	FULL THROTTLE			
12000								
FOR MAX TAKEOFF POWER (5 MIN), ADD 2.8 IN.								

RAVEN II
NEVER EXCEED SPEED - KIAS

PRESS	2200 LB TOGW & BELOW							
	OAT - °C							
ALT-FT	-30	-20	-10	0	10	20	30	40
SL								
2000	130						127	123
4000					126	122	118	114
6000			126	122	117	113	108	103
8000	126	122	117	112	107	101	96	91
10000	117	112	106	101	95	90	85	
12000	107	101	95	89	NO FLIGHT			
14000	95	89						
OVER 2200 LB TOGW, SUBSTRACT 10 KIAS								
FOR AUTOROTATION, SUBSTRACT 30 KIAS								

ASTRO + RAVEN I
LIMIT MANIFOLD PRESSURE - IN. HG

PRESS	MAXIMUM CONTINOUS POWER						
	OAT - °C						
ALT-FT	-20	-10	0	10	20	30	40
SL	22.9	23.2	23.5	23.8	24.1	24.4	24.7
2000	22.5	22.8	23.1	23.4	23.7	24.0	24.2
4000	22.2	22.5	22.8	23.1	23.4	23.7	23.9
6000	21.8	22.1	FULL THROTTLE				
FOR MAX TAKEOFF POWER (5 MIN), ADD 1.6 IN.							

ASTRO + RAVEN I
NEVER EXCEED SPEED - KIAS

PRESS	2200 LB TOGW & BELOW						
	OAT - °C						
ALT-FT	-20	-10	0	10	20	30	40
SL							
2000	130					127	123
4000				126	122	118	114
6000		126	122	117	113	108	103
8000	122	117	112	107	101	96	91
10000	112	106	101	95	90	85	
12000	101	95	89	NO FLIGHT			
14000	89						
OVER 2200 LB TOGW, SUBSTRACT 10 KIAS							
FOR AUTOROTATION, SUBSTRACT 30 KIAS							

CLIPPER - WITH FLOATS
NEVER EXCEED SPEED - KIAS

2200 LB TOGW & BELOW							
PRESS	OAT - °C						
ALT-FT	-30	-20	-10	0	10	20	30	40
SL								
2000	120						117	113
4000					116	112	108	104
6000			116	112	107	103	98	93
8000	116	112	107	102	97	91	86	81
10000	107	102	96	91	85	80	75	
12000	97	91	85	79	NO FLIGHT			
14000	85	79						

OVER 2200 LB TOGW, SUBSTRACT 10 KIAS
FOR AUTOROTATION, SUBSTRACT 20 KIAS

CLIPPER - WITHOUT FLOATS
NEVER EXCEED SPEED - KIAS

2200 LB TOGW & BELOW							
PRESS	OAT - °C						
ALT-FT	-30	-20	-10	0	10	20	30	40
SL								
2000	130						127	123
4000					126	122	118	114
6000			126	122	117	113	108	103
8000	126	122	117	112	107	101	96	91
10000	117	112	106	101	95	90	85	
12000	107	101	95	89	NO FLIGHT			
14000	95	89						

OVER 2200 LB TOGW, SUBSTRACT 10 KIAS
FOR AUTOROTATION, SUBSTRACT 30 KIAS

Chapter 3

Emergency Procedures

Chapter 3
Emergency Procedures

Definitions

Land Immediately: Land on the nearest clear area where a safe normal landing can be performed. Be prepared to enter an autorotation during the approach, if required.

Land as soon as practical: Land at the nearest airport or other facility, where emergency maintenance may be performed.

Maximum Glide Distance Configuration

- Airspeed approximately 90 KIAS
- Rotor RPM approximately 90 %
- Best glide ratio is about 4.7 : 1 or one nautical mile per 1300 feet AGL

Minimum Rate Of Descent Configuration

- Airspeed approximately 55 KIAS
- Rotor RPM approximately 90 %
- Minimum rate of descent is about 1350 feet per minute. Glide ratio is about 4 : 1 or one nautical mile per 1500 feet AGL.

Caution:
Increase the rotor RPM to 97 % minimum, when autorotating below 500 feet AGL.

Power Failure In General

A power failure may be caused by either an engine or drive system failure and will usually be indicated by the low RPM horn.

An engine failure may be indicated by a change in noise level, a left yaw of the helicopter's nose, illumination of the OIL pressure warning light or decreasing engine RPM.

A drive system failure may be indicated by an unusual noise or vibration, a right or left yaw of the helicopter's nose or a decreasing rotor RPM, while the engine RPM is increasing.

Allow airspeed to reduce to power-off V_{ne} or below.

Caution:
Aft cyclic is required when collective is lowered at high speed and forward CG.

Caution:
Avoid using aft cyclic during touchdown or during ground slide to prevent possible blade strike to tailcone.

Power Failure Above 500 Feet AGL

- Lower the collective immediately to maintain RPM and enter a normal autorotation.

- Establish a steady glide at approximately 70 KIAS (See "Maximum Glide Distance Configuration", page 3-3)

Emergency Procedures

- Adjust the collective to keep the RPM in the green arc or apply full down collective if light weight prevents attaining a RPM above 97 %.

- Select a landing spot and, if altitude permits, maneuver so that a landing can be established into the wind.

- A restart may be attempted at pilot's discretion if sufficient time is available (See "Air Restart Procedure" below).

- If unable to restart, turn off unnecessary switches and shut off the fuel.

- At about 40 feet AGL begin a cyclic flare to reduce the rate of descent, as well as the forward speed.

- At about 8 feet AGL apply forward cyclic to level the helicopter and raise collective just before touchdown to cushion the landing.
  Touch down in a level attitude with the nose of the helicopter pointing straight ahead.

Note:
If power failure occurs at night do not turn on the landing lights above 1000 feet AGL to preserve battery power.

Power Failure Between 8 Feet And 500 Feet AGL

- Takeoff Operation should be conducted per the Height-Velocity Diagram, where shaded areas should be avoided.

- If a power failure occurs, lower the collective immediately to maintain rotor RPM.

- Then adjust the collective to keep RPM in the green arc or apply full down collective if light weight prevents attaining a RPM above 97 %.

- Maintain airspeed until ground is approached, then begin a cyclic flare to reduce the rate of descent, as well as the forward speed.

- At about 8 feet AGL apply forward cyclic to level the helicopter and raise collective just before touchdown to cushion the landing. Touch down in a level attitude with the nose of the helicopter pointing straight ahead.

Power Failure Below 8 Feet AGL

- Apply right pedal as required to prevent the helicopter from yawing.

- Allow the aircraft to settle, and then

- Raise the collective just before touchdown to cushion the landing.

Air Restart Procedure

Caution:
Do not attempt an air restart if an engine
malfunction is suspected or before a safe
autorotation is established. Air restarts are not
recommended below 2000 feet AGL.

<u>Astro and Raven I</u>

- Mixture – full rich
- Throttle – closed, then cracked slightly
- Actuate the starter with the left hand

<u>Raven II</u>

- Mixture – Off
- Throttle – Closed
- Starter – Engage
- Mixture – Move slowly rich, while cranking

Ditching To Water – Power Off

- Follow the same procedure as for power failure over land until contacting the water.

- Then apply lateral cyclic when the aircraft contacts water to stop the blades from rotating.

- Release the seat belt and quickly clear the aircraft, when the blades stop rotating.

Ditching To Water – Power On

- Descend into a hover above the water, then

- Unlatch the doors and let the passengers exit the helicopter.

- Fly to a safe distance from the passengers to avoid possible injury by the blades.

- Switch off battery and alternator, then roll the throttle into detent spring.

- Keep the aircraft level and apply full collective as the aircraft contacts the water.

- Then apply lateral cyclic when the aircraft contacts water to stop the blades from rotating.

- Release the seat belt and quickly clear the aircraft, when the blades stop rotating.

Loss Of Tail Rotor Thrust During Forward Flight

- A tail rotor failure is usually indicated by a right yaw of the helicopter's nose, which cannot be corrected by applying left pedal.

- Immediately enter an autorotation and

- Maintain at least 70 KIAS if practical.

- Select a landing site, roll off the throttle into detent spring and perform an autorotation landing.

Note:
When a suitable landing site is not available, the vertical fin may permit limited controlled flight at low power settings and airspeeds above 70 KIAS; however, prior to reducing the airspeed, re-enter a full autorotation.

Loss Of Tail Rotor Thrust During Hover

- A tail rotor failure is usually indicated by a right yaw of the helicopter, which cannot be stopped by applying left pedal.

- Immediately roll the throttle off into detent spring and allow the aircraft to settle.

- Raise the collective just before touchdown to cushion the landing.

Engine Fire During Start On Ground

- If cranking – continue and attempt to start, which would suck flames and excess fuel into the engine.

- If the engine starts, run it at 60-70 % RPM for a short time, then shut down and inspect for damage.

- If the engine fails to start, shut off fuel and the master battery switch.

- Extinguish the fire with a fire extinguisher, wool blanket or dirt.

- Then inspect the helicopter for damage.

Fire In Flight

- Enter an autorotation and,

- If time permits, switch off
  - the master battery switch
  - the cabin heat

  and open the cabin vent to establish air circulation.

- If the engine is running perform a normal landing and immediately shut off the fuel valve.

- If the engine stops running shut off the fuel valve and execute a normal autorotation landing.

Electrical Fire In Flight

- Switch off
  - the master battery
  - the alternator switch

- Land immediately

- Extinguish the fire and inspect the helicopter for damage.

Caution:
The Low RPM warning system, as well as the governor are inoperative with the master battery and the alternator switch both being in the off position.

Tachometer Failure

If the rotor or the engine tachometer malfunctions in flight, use the remaining tachometer to monitor the RPM.
If it is not clear, which tachometer is malfunctioning or if both tachometers malfunction, allow the governor to control the RPM and land as soon as practical.

Note:
Each tachometer, the governor and the low RPM warning system are on separate circuits. A special circuit allows the battery to supply power to the tachometers, even if the master battery and alternator switch are both off.

Cyclic Trim Failure
(If Hydraulics Not Installed)

If the automatic cyclic trim fails to compensate for the cyclic feedback forces, land as soon as practical.
If the trim generates undesirable forces, switch trim off and land as soon as practical.

Hydraulic System Failure
(If Hydraulics Installed)

A hydraulic system failure is indicated by heavy or stiff cyclic and collective controls. Loss of hydraulic fluid may cause intermittent and/or vibrating feedback in the controls.
Control will be normal, except for the increase in stick forces.

• Adjust the airspeed and flight condition as desired for comfortable control.

• Verify the hydraulics switch is selected ON.

• If the hydraulics are not restored switch the hydraulics OFF, using the switch.

• Land as soon as practical.

Governor Failure

If the engine RPM governor malfunctions, grip the throttle firmly to override the governor and then switch the governor off. Complete the flight using manual throttle control.

Warning And Caution Lights

Note:
Some warning lights may not be found in earlier helicopters, check
which warning lights are applicable to the specific helicopter you fly.

Note:
If a light causes excessive glare at night, the bulb may be unscrewed
or the circuit breaker pulled to eliminate irritating glare during the
landing.

OIL

Indicates loss of engine power or oil pressure.

• Check the engine tachometer for power loss.
• Check the oil pressure gage and, if a pressure loss is confirmed,
  land immediately.

Continued operation without sufficient oil pressure will cause serious
engine damage and an engine failure may occur.

ENG FIRE

Indicates a possible fire in the engine compartment.

• See the emergency procedures regarding engine fire.

MR TEMP

Indicates an excessive temperature of the main rotor gearbox.

• See opposite note.

MR CHIP
Indicates metallic particles in the main rotor gearbox.

• See note below.

TR CHIP
Indicates metallic particles in the tail rotor gearbox.

• See note below.

Note:
If the light is accompanied by any indication of a problem, such
as noise vibration or a temperature rise, land immediately.
If there is no other indication of a problem,
land as soon as practical.

Break-in fuzz will occasionally activate the chip lights.
If no metal chips or slivers are found on the detector plug, clean
and reinstall (tail rotor gearbox must be refilled with
new oil). Afterwards hover for at least 30 minutes.
If the chip light comes on again, replace the gearbox before
further flight.

LOW FUEL
Indicates approximately three gallons of useable fuel are remaining. The engine will run out of fuel after ten minutes at cruise power.

Caution:
Do not use the low fuel warning light as a
working indication of fuel quantity.

AUX FUEL PUMP

Indicates low auxiliary fuel pump pressure.

- If no other indication of a problem, land as soon as practical.
- If the light is accompanied by erratic engine operation, land immediately.

FUEL FILTER

Indicates a fuel strainer contamination.

- If no other indication of a problem, land as soon as practical.
- If the light is accompanied by the AUX FUEL PUMP warning light-tor erratic engine operation, land immediately.

CLUTCH

Indicates that the clutch actuator circuit is on, either engaging or disengaging the clutch.
When the switch is in the engage position, the light stays on until the belts are properly tensioned.

- Never take off before the light goes out.

Note:

The CLUTCH light may come on momentarily during run-up or during flight to retension the belts, as they warm up and stretch slightly. This is normal. If, however, the light flickers or comes on in flight and does not go out within 7 or 8 seconds, pull the CLUTCH circuit breaker, reduce the power and land immediately.
Be prepared to enter an autorotation.
Inspect the drive system for a possible malfunction.

ALT
Indicates a low voltage and a possible alternator failure.

- Turn off nonessential electrical equipment and switch ALT off and back on after one second to reset the overvoltage relay.
- If the light stays on, land as soon as practical.

Continued flight without a functioning alternator can result in the loss of the electronic tachometer, producing a hazardous flight condition.

BRAKE
Indicates that the rotor brake is engaged.

- Release immediately in flight or before starting the engine.

STARTER-ON
Indicates that the starter motor is engaged.

- If the light does not go out when the starter button is released, immediately pull the mixture to the idle cut-off and turn the master switch off.
- Have the starter motor serviced.

GOV OFF
Indicates that the engine RPM throttle governor is off.

CARBON MONOXIDE

Indicates elevated levels of carbon monoxide (CO) inside the cabin.

- Open the nose and door vents and shut off the heater.
- If hovering, transition to forward flight.
- If symptoms of CO poisoning (headache, drowsiness, dizziness) accompany the light, land immediately.

LOW RPM HORN & CAUTION LIGHT

A horn and an illuminated caution light indicate that the rotor RPM may be below safe limits.

- To restore the RPM, immediately roll throttle on, lower collective and, in forward flight, apply aft cyclic.

The horn and the caution light are disabled when the collective is full down.

Chapter 4

Normal Procedures

Chapter 4
Normal Procedures

Airspeeds for safe operation

Takeoff and Climbs	60 KIAS
Maximum Rate of Climb (V_y)	55 KIAS
Maximum Range	100 KIAS *
Landing Approach	60 KIAS
Autorotation	70 KIAS *

*Certain conditions may require lower airspeeds.
 See never exceed speed table.

Hover and Takeoff Procedure

Verify that the doors are latched.
The doors should all be latched but unlocked to allow rescue or exit in an emergency.
Verify that the governor, as well as the hydraulics are switched on. Verify the RPM is stabilized at 101 to 102 %.

Clear the area. Open the door, if necessary, and verify no person or obstacle may interfere with the helicopter.
Slowly raise the collective until the aircraft is light on the skids. Then reposition the cyclic as required for equilibrium (a good help for finding this position is to look at a fix point out and ahead of the helicopter, which improves the feeling for the helicopter's movement) and, if necessary, cancel yawing of the helicopter using the pedals.
Then gently lift the aircraft into a hover by raising collective further.
Be cautious and ready for corrective control inputs that may be required lifting the helicopter up into a hover.

Once the helicopter is in a hover check the gages and verify they are in the green operating areas.

When hovering caution should be used over areas with loose surface (e.g. sand, dust or snow), as brown- or white-out situations may occur due to the rotor downwash and dramatically reduce the pilot's visibility.
Another factor to consider is the effect the helicopter's downwash can have on persons and objects at close range. Therefore hovering in close proximity of people or bigger objects (e.g. parked airplanes) should be avoided.

To takeoff out of the hover lower the nose of the helicopter using the cyclic and accelerate to climb speed following the profile shown by the height-velocity diagram shown in the Loading and Performance section of this book.
Avoid exceeding two inches of MAP above the IGE hover power to prevent an excessive nose-down attitude.
If the RPM drops below 101% lower the collective.

Cruise

Once in cruise verify that the RPM is within the green arc.
Then set a manifold pressure with the collective that is appropriate for the flight profile to be flown. For each airspeed setting
there is a manifold pressure setting, which will keep the helicopter in level flight without any change in altitude.
A technique often used in straight-and-level flying is to align an
object inside the cockpit (e.g. the compass) with the horizon and
use it as a reference point.
Verify that all gages are in the green and all warning lights are
out.

Caution:
As opposed to most airplanes the mixture is <u>not</u>
leaned for cruise flight in the R44. The mixture
must be full rich during flight.

Note:
Slight yaw oscillation during cruise can be stopped
by applying a small amount of pedal.

Normal Procedures

Approach and Landing

Perform the final approach into the wind at the lowest practical rate of descent and with an initial airspeed of 60 knots.
Reduce the airspeed and altitude smoothly and gradually until in a hover. Doing so always make sure that the rate of descent is less than 300 feet per minute before reducing the airspeed to below 30 KIAS.

From the hover lower the collective gradually until ground contact. Once again focusing on a fix object outside and in front of the helicopter helps maintain a steady helicopter attitude while setting down.
After initial ground contact lower the collective all the way to the full down position.
Use the Shutdown Checklist to safely shut down the helicopter.

Note that the helicopter should always touch the ground in a vertical or slight forward movement. Avoid touching down with sideward or rearward movement, as this might lead to dynamic rollover of the helicopter.

Caution:
When landing on a slope, return the cyclic control to the neutral position before final reduction of the rotor RPM.

Caution:
Never leave the helicopter flight controls unattended while the engine is running.

Doors-Off Operation

There are multiple reasons for flying with one or more doors removed.
It should be kept in mind however that Doors-Off operation is bound to limitations. For example the maximum airspeed allowed decreases to 100 KIAS with any door(s) removed.

Passengers should be warned to keep head and arms inside the cabin, as well as to securely stow any loose objects to avoid damage or injury due to the high-velocity airstream around the fuselage.

Removing the left side doors should be avoided due to the high risk of loose objects hitting and severely damaging the tail rotor. Even small objects can, accelerated by the high-velocity air-stream, cause great danger.

Baggage should not be placed underneath the rear seats if not occupied during doors-off operation since it could cause the rear seat bottoms to lift and items to be blown out.
Also all seat belts should be fastened during Doors-Off operations.

Hydraulics-Off Training

To simulate hydraulic system failure the cyclic-mounted hydraulic switch is used.
Caution should be used when switching the hydraulics back on after the training is concluded. The force used on the cyclic and collective should be relaxed before switching from off to on, to avoid overcontrolling.

Governor-Off Training

To simulate governor system failure the governor is selected off by using the switch located on the pilot's collective.
Be cautious to switch the governor back on after the training session is concluded, as inadvertent governor-off operations could lead to an accident. Also flight with the governor selected off is not allowed in the R44, except for governor-off training or inflight governor malfunction.

Autorotation

An autorotation is a descending maneuver where the engine is disengaged from the main rotor system and the rotor blades are driven solely by the upward flow of air through the rotor. In other words, the engine is no longer supplying power to the main rotor.

The most common reason for an autorotation is an engine failure, but autorotations can also be performed in the event of a complete tail rotor failure, since there is virtually no torque produced in an autorotation or, if altitude permits, they can also be used to recover from settling with power.

This is possible because of a freewheeling unit, the sprag clutch, which connects the engine to the drive system and automatically disengages the engine from the rotor system, allowing it to rotate freely anytime the engine RPM is less than the rotor RPM.

The following description is valid for autorotations, performed for training purposes only. Refer to the Emergency Procedures section for details about engine failure emergency procedures.

Normal
Procedures

Practice Autorotation with Power Recovery

• Lower the collective to the down stop and adjust the throttle as required for a small tachometer needle separation.

Caution:
To avoid inadvertent engine stoppage DO NOT roll throttle to full idle. Roll throttle off smoothly enough for a small visible needle split.

Note:
The governor is not active below 80% engine RPM regardless of the governor switch position.

Note:
When entering autorotation from above 6000 feet, reduce throttle slightly before lowering collective to prevent engine overspeed.

• Raise collective as required to keep the rotor RPM from going above the green arc and adjust the throttle for small needle separation.

• Keep the RPM in the green arc and the airspeed between 60 and 70 KIAS.

• At about 40 feet AGL, begin cyclic flare to reduce the rate of descent and the forward speed.

• At about 8 feet AGL, apply forward cyclic to level the aircraft and raise collective to control the descent. Add throttle if required to keep RPM in green arc.

Full-down autorotations (all the way to the ground), if required to perform for demonstrational purposes, are performed in the same way as a power recovery autorotation, except that:

Prior to the cyclic flare throttle must be rolled off into the detent spring and be hold against the hard stop until the autorotation maneuver is complete.
This is because if the throttle would not be rolled into detent the throttle correlator would automatically add power, as soon as the collective is raised in the end of the maneuver.

ALWAYS CONTACT THE GROUND WITH THE SKIDS LEVEL AND THE NOSE HEADED STRAIGHT AHEAD

Caution:
During simulated engine failures, rapid decrease in rotor RPM will occur, requiring immediate lowering of the collective to avoid dangerously low rotor RPM. Catastrophic rotor stall could occur if rotor RPM ever drops below 80% plus 1% per 1000 feet of altitude.

Note:
When practice autorotations are performed with ground contact rapid wear of the landing skid shoes occurs. Inspect the skid shoes periodically and replace them, when the minimum shoe thickness is 0.06 inches (1.5 mm).

Parking

When parking the helicopter the cyclic stick should be placed in the
neutral position and the friction be applied.
Then the collective should be put full down and the respective friction
should be applied as well.
The rotor blades should be aligned with the centerline of the helicopter's fuselage. If windy conditions are encountered or expected the
blades should be aligned slightly offset to prevent the blade from flapping into the tailcone.
Then the rotor brake should be applied.
During storm conditions the helicopter should be hangered or moved
to a safe area.

Pushing in the locking pins located on each door locks the rear
doors, while the front doors are locked using key locks.

Ground Handling

There are different ways to move the R44 on the ground.
Probably the most common method is to use ground-handling
wheels that are provided with each new helicopter.
For this 2 people are required in order to move the helicopter.
One is holding down on the tail rotor gearbox, lifting the front
skids of the helicopter off the ground and at the same time steer-
ing the direction of movement.
It is important not to apply force to any other part of the tail except the gearbox to prevent damage.
The second person will push the fuselage. The steel tube frame
inside the aft cowl door may be used for pushing.
Ground Handling is best done by having the blades aligned with
the helicopter's longitudinal axis, which allows for better judging
of distances to obstacles. The rotor brake can be
applied.

Another method is using an electric tow cart that can be
purchased from Robinson Helicopter Company. For usage see the
instructions that are provided by the manufacturer.

Lastly there is the method of having a pad that can be moved.
The helicopter lands on the pad that then may be moved either
by hand or by using a towing truck.

Normal
Procedures

Chapter 5

Checklists

Checklists

Chapter 5
Checklists

Basic Checks

- Remove any temporary covers, tie downs and wheels, if attached
- In cold ambient conditions remove frost, ice or snow, if found, even if only small amounts
- Check the maintenance records to verify the aircraft is airworthy
- Check the general condition of the aircraft
- Verify there are no leaks
- Verify there is no discoloration due to heat
- Verify there are no dents, chafing, galling, nicks, corrosion or cracks
- Verify there is no fretting at seams where parts are joined together (Fretting of aluminum parts produces a fine black powder while fretting of steel parts produces a reddish brown or black residue)
- Verify the Telatemps do not show unexplained temperature increases

An 8-foot ladder is recommended for the preflight inspection of the main rotor. The main rotor hub however may also be reached by opening the right rear seat and the right forward cowl door and then first stepping on the seat support and then on the deck below the aux fuel tank.

An advantage of using the ladder is that the main rotor blades may be inspected more carefully.

Caution:
Do not pull blades down, as damage may occur. To lower one blade, push the opposite blade up.

Note: Numbers in parentheses indicate the part's location on the pictures.

Checklist Pictures

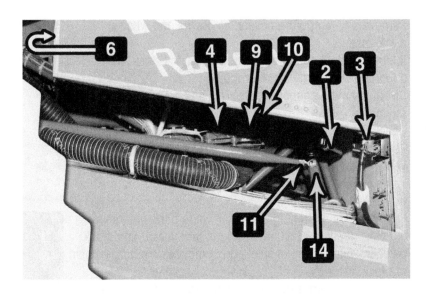

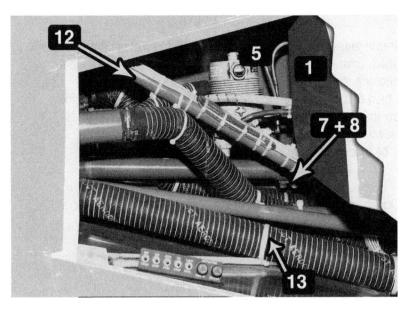

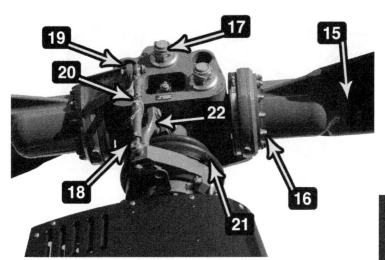

Checklists

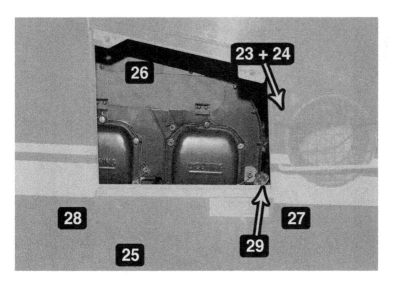

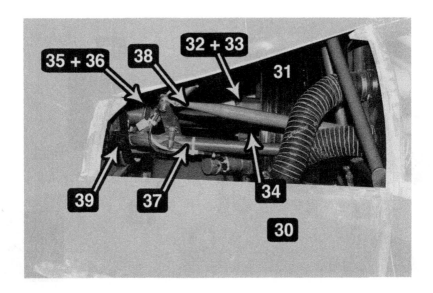

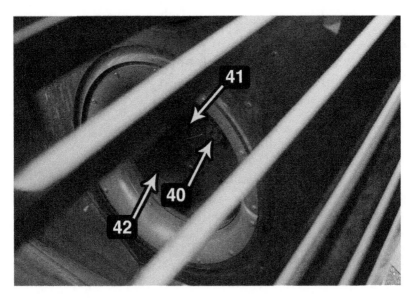

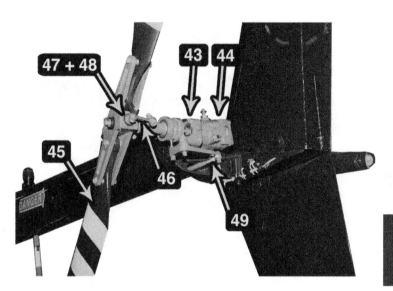

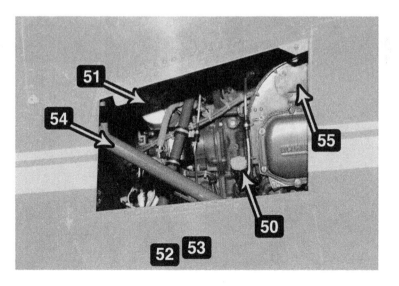

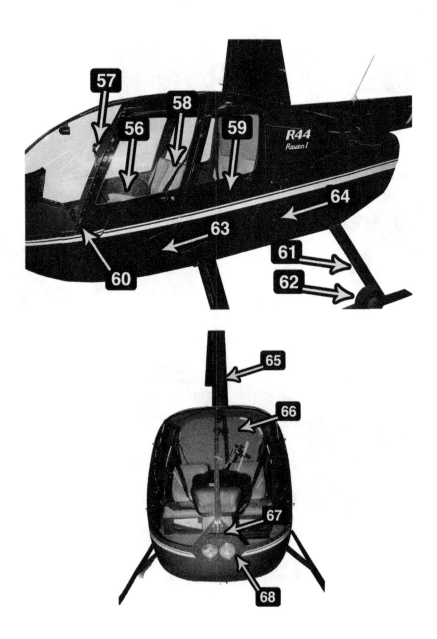

Preflight

Cabin Interior	
Required Documents (varying per country)	On Board
Master Switch	On
Clutch Switch	On
Oil Pressure, ALT, CLUTCH Lights	Illuminated
Governor Light (if GOV switch set to OFF)	Illuminated
AUX Fuel Pump Light (Raven II only)	Illuminated
LAND LTS, NAV LTS, STROBE Switch	On
Landing, Position, Strobe Lights	Check functioning
Warning Light Test Switches	Push to Test
Clutch Switch	Off
Fuel Quantity	Check Gages + Visual Check (Order Fuel if necessary)
CLUTCH Light	Not illuminated
Master Switch	Off
Clock	Functioning
Instruments, Switches and Controls	Check Condition
Loose Articles	Removed or stowed
Baggage Compartments (all)	Check
Seat Belts	Check condition and fastened
Adjustable Pedals	Pins secure

Checklists

Upper Forward Cowl Doors - Right Side	
Fuel Filler Cap	Tight
Aux Fuel Tank (1)	No leaks
Fuel Lines (2)	No leaks
Fuel Tank Sump, Gascolator Drains (3)	Sample
Gearbox Oil (4)	Full, no leaks
Hydraulic System (if Hydraulics installed) (5)	Fluid full, no leaks
Rotor Brake (6)	Actuation normal
Flex Coupling (7)	No cracks, nuts tight
Yoke Flanges (8)	No cracks
Gearbox Telatemp (9)	Normal
Hydr. Pump Telatemp (if Hydraulics installed) (10)	Normal
Control Rod Ends (11)	Free without looseness
Steel Tube Frame (12)	No cracks
All Fasteners (13)	Tight
Tail Rotor Control (14)	No interference

Main Rotor	

Caution:
Do not pull rotor blades down, as damage may
occur. To lower one blade, push the opposite blade up.

Blades (15)	Clean, no damage or cracks

Caution:
Verify erosion on the lower surface of the blades
has not exposed the skin-to-spar bond line.

Pitch Change Boots (16)	No leaks
Main Hinge Bolts (17)	Cotter Pins installed
All Rod Ends (18)	Free without looseness
Pitch Link Jam Nuts (19)	Tight
Pitch Link Safety Wire (20)	Secure
All Fasteners (21)	Tight
Swash plate Scissors (22)	No excessive looseness
Upper forward Cowl Doors	Latched

Checklists

Lower Cowl Door - Right Side	
Air Box and Duct (Raven II only) (23)	Secure
Carburetor Air ducts (Astro and Raven I) (24)	Secure
Carburetor Heat Scoop (Astro and Raven I) (25)	Secure
Engine Sheet Metal (26)	No cracks
Fuel Lines (27)	No leaks
Oil Lines (28)	No leaks or chafing
Exhaust System	No cracks
Throttle Linkage (Raven II - others Left Cowl Door)	Operable
Primer (if installed) (29)	Prime as required/ Locked/ No Leaks
Cowl Door	Latched

Aft Cowl Door - Right Side	
Oil Cooler Door (30)	Check
V-Belt Condition (31)	Check
V-Belt Slack	Check
Sprag Clutch (32)	No leaks
Upper Bearing (33)	No leaks
Telatemp - upper Bearing (34)	Normal
Sheave Condition	Check
Flex Coupling (35)	No cracks, nuts tight
Yoke Flanges (36)	No cracks
Steel Tube Frame (37)	No cracks
Tail Rotor Control (38)	No interference
Tailcone Attachment Bolts (39)	Check
Cowl Door	Latched

Checklists

Engine Rear	
Cooling Fan Nut (40)	Pin in line with Marks
Cooling Fan (41)	No cracks
Fan Scroll (42)	No cracks
Tailpipe Hanger	No cracks

Empennage	
Tail Surfaces	No cracks
Fasteners	Tight
Position Light	Check
Tail Rotor Guard	No cracks

Tail Rotor	
Gearbox Telatemp (43)	Normal
Gearbox (44)	Oil visible, no leaks
Blades (45)	Clean and no damage or cracks
Pitch Links (46)	No looseness
Teeter Bearings (47)	Check Condition
Teeter Bearing Bolt (48)	Does not rotate
Control Bell crank (49)	Free without looseness

Tailcone	
Rivets	Tight
Skins	No cracks or dents
Strobe Light Condition	Check
Antenna	Check

Cowl Door - Left Side	
Engine Oil (50)	7-9 qts
Oil Filter (51)	Secure, no leaks
Throttle Linkage (Astro and Raven I) (52)	Operable
Battery and Relay (if located here) (53)	Secure
Steel Tube Frame (54)	No cracks
Engine Sheet Metal (55)	No cracks
Exhaust System	No cracks
Cowl Door	Latched

Fuel Tank (Main)	
Filler Cap	Tight
Leakage	None
Sump Drain (Non-Bladder Tank)	Sample

Checklists

Fuselage - Left Side	
Baggage Compartments (56)	Check
Removable Controls (if installed) (57)	Secure
Collective Control	Clear
Seat Belts (58)	Check condition and fastened
Doors (59)	Unlocked and latched
Door Hinge Safety Pins (60)	Installed
Landing Gear (61)	Check
Ground Handling Wheel (62)	Removed
Position Light (63)	Check
Static Port (64)	Clear

Nose Section	
Pitot Tube (65)	Clear
Windshield Condition and Cleanliness (66)	Check
Fresh Air Vent (67)	Clear
Landing Lights (68)	Check

Pop-Out Floats (if installed)	
Float and Float Cover Condition	Check
Hose and Fitting Condition	Check
Pressure Cylinder	Check Pressure (see POH)
Safety Pin at Pressure Cylinder	Verify Removed
Inflation Lever Safety	"Ready" or "Locked" as required

Inflatable Floats (if installed)	
Float Pressure	Check (see POH)
Float Condition	Check

Caution:
Remove the left seat controls if the person in that seat is not a rated helicopter pilot.

Caution:
Fill the baggage compartments under unoccupied seats to capacity, before using baggage compartments under occupied seats. Avoid placing objects in the compartment, which would injure the occupant if the seat collapses during a hard landing.

Caution:
Ensure that all doors are unlocked before flight, to allow for rescue or exit in an emergency.

Caution:
Shorter pilots may require cushion to obtain full travel of all controls. When using the cushion, verify that aft cyclic travel is not restricted.

Caution:
Be sure that the rotor blades are approximately level to avoid possible tailcone strike.

Checklists

Before Starting Engine

Seat Belts	Fastened
Fuel Shut-Off Valve	On
Cyclic Friction	Off
Collective Friction	Off
Cyclic, Collective, Pedals	Full travel free
Throttle	Full travel free
Collective	Full down, Friction on
Cyclic	Neutral, Friction on
Pedals	Neutral
Trim Switch (if no Hydraulics installed)	Off
Landing Light Switch	Off
HYD Switch (if Hydraulics installed)	On
Governor Switch	On
Circuit Breakers	In

Alpha and Raven I only	
Carb Heat	Off
Mixture	Full rich
Mixture Guard	Installed

Clutch	Disengaged
Altimeter	Set
Rotor Brake	Disengaged

Engine Starting Tips (Raven II only)

During priming, the aux fuel pump warning light may remain illuminated momentarily. Continue priming for 3 to 5 seconds after the light extinguishes.

If the engine does not fire after 5 to 7 seconds of cranking, repeat the priming sequence and reattempt start. If the engine fails to start after three attempts, allow the starter to cool for ten minutes before the next attempt.

If the engine fires momentarily but dies before or while moving the mixture to rich, pull the mixture to off, engage the starter and then push mixture slowly rich, while cranking.

Checklists

Starting Engine and Run-Up

Throttle twists for priming	As required
Throttle	Closed
Master Switch	On
Area	Clear
Strobe Light	On

Astro and Raven I	
Ignition Switch	Start, then Both

Raven II	
Mixture	Rich
Ignition Switch	Prime, then Both
Mixture	Pull off
Starter	Engage until engine fires
Mixture	Move full rich
Mixture Guard	Installed

Starter-On light	Out
Set Engine RPM	50 to 60 %
Clutch Switch	Engaged
Blades turning	Less than 5 seconds
Alternator Switch	On
Oil Pressure within 30 seconds	25 psi minimum
Avionics, Headset	On
Wait for Clutch Light	Out
Warm-Up RPM	60 to 70 %

Engine Gages	Green
Mag Drop at 75% RPM	7 % max in 2 sec
Carburetor Heat Check (Astro and Raven I)	CAT rise/drop
Sprag Clutch Check at 75% RPM	Needles split
Doors	Closed and latched
Limit MAP Chart	Check
Cyclic Friction	Off
Collective Friction	Off
Hydraulic System (if Hydraulics installed)	Check
Cyclic Trim (if no Hydraulics installed)	Check
Governor On, increase Throttle	RPM 101-102 %
Warning Lights	Out
Lift Collective slightly, reduce RPM	Horn/Light at 97 %

Caution:
On slippery surfaces, be prepared to counter a nose right rotation with left pedal as the governor increases the RPM.

Continued on next page.

Note:
For trim check, switch trim on and verify balanced cyclic forces.
For a hydraulic system check, use small cyclic inputs.
With hydraulics off, there should be approximately one half inch
of freeplay before encountering control stiffness and feedback.
With hydraulics on, the controls should be free with no feedback or
uncommanded motion.

Note:
During run-up and shutdown, the pilot should uncover his/her right
ear, open the right door and listen for unusual bearing noise.
Failing bearings will produce an audible whine or growl well before
final failure.

Note:
Idle mixture and speed may require adjustment as conditions vary
from sea level standard.
Refer to the R44 Maintenance Manual for idle adjustment procedures.

Engine Shut-Down

Collective down, 60 - 70 % RPM	Collective Friction On
Cyclic and Pedals neutral	Cyclic Friction On
CHT drop	Recommended 2 minutes and 300 °F
Throttle	Closed
CLUTCH Switch	Disengage
Avionics, NAV LTS Switch	Off
Wait 30 seconds	Mixture Off
ALT, MAGNETOS	Off
Wait 30 seconds	Apply Rotor Brake
CLUTCH Light Off	MASTER Switch Off

Caution:
Do not slow the rotor by raising collective during the
shutdown. The blades may flap and strike the tailcone.

Note:
HYD switch should be left on for start-up and shutdown to reduce
battery drain and the possibility of unintentional hydraulics-off
liftoff. Switch off only for pre-takeoff controls check or hydraulics-
off training.

Note:
The rotor brake should be left engaged after shutdown to
disable the starter buttons and reduce the possibility of
unintentional starter engagement.

Checklists

Chapter 6

Loading & Performance

Chapter 6

Loading & Performance

The engineers designing a helicopter determine the amount of cyclic control power that is available, and establish both longitudinal and lateral Center of Gravity (CG) envelopes that are used by the pilot to load the helicopter in a manner allowing for sufficient cyclic control during all flight conditions.

If the CG is ahead of the forward limit, the helicopter will tilt, the rotor disk will have a forward pull and rearward cyclic is required to counteract this.

If the CG is too far forward, there may not be enough cyclic authority to allow the helicopter to flare for a landing, and it will consequently require an excessive landing distance.

If the CG is aft of the allowable limits on the other hand, the helicopter will tend to fly with a tail-low attitude and may need more forward cyclic stick displacement than is available to maintain a hover in a no-wind condition. Resulting there might not be enough cyclic power to prevent the tail boom striking the ground or, if gusty winds should cause the helicopter to pitch up during high-speed flight, to lower the nose.

Helicopters are approved for a specific maximum gross weight, 2400 lbs. for Astros and Raven I's and 2500 lbs. for Raven II's respectively, however it is not safe to operate them at this weight under all conditions.

A high-density altitude decreases the safe maximum weight as it affects the hovering, takeoff, climb, autorotation, and landing performance.

Since the fuel tanks are located behind the CG, the helicopter tends to shift forward as fuel is used. Under some flight conditions, the balance may shift enough that there will not be sufficient cyclic authority to flare for landing.

Therefore the loaded CG should be computed for both takeoff and landing weights.

The weights recorded in the respective weight and balance record are added and their distances from the datum are used to compute the moments at each weighing point.

The total moment is divided by the total weight to determine the location of the CG in inches from the datum.

The datum of some helicopters, as in the R44, is located ahead of the aircraft resulting in all longitudinal arms being positive.

The lateral CG is determined in the same way as the longitudinal CG, except the distances between the scales and the lateral centerline are used as the arms. Here, arms to the right of the centerline are positive and those to the left are negative for calculation purposes.

The centerline is a line through the symmetrical center of an aircraft from nose to tail and serves as the datum for measuring the arms used to find the lateral CG.

Lateral moments that cause the aircraft to rotate clockwise are positive (+), and those that cause it to rotate counter-clockwise are negative (-).

Important to remember is that the Weight and Balance records are different for each single helicopter and therefore the weights and arms provided in the specific helicopter's flight manual must be used for calculations.

Different terms are used to describe specific weights in aviation.
The ones used most are described below:

Basic empty weight
This includes the weight of the:
- Standard helicopter
- Optional equipment
- Unusable fuel
- Full operation fluids (include full engine oil)

Useful load
This includes the weight of the:
- Flight Crew
- Passengers
- Cargo or baggage

Gross weight
This is the sum of empty weight and useful load and therefore
includes all of the above. There is a maximum gross weight,
as well as a minimum gross weight.
For flight the gross weight has to be between these two values.

This has multiple reasons.
Firstly the CG will be out of the limits and sufficient control avail-
ability may not be given.
Furthermore if the gross weight is too high the helicopter will not
have sufficient power and the airframe might not be able
to withstand the then acting forces.

If the gross weight is too low on the other hand there might not
be enough weight to gain and maintain sufficient rotor RPM in an
autorotation.

Loading &
Performance

Methods of Weight and Balance calculation

There are different methods that can be used to calculate weight and balance of a helicopter. The two most used for the R44, the mathematical and the graphical, will be explained in the following.

In the mathematical method all weights are added up to get the Gross Weight, which should be within the weight limits as described in the Flight Manual.
Each weight item is then multiplied with its relating arm to get the moment and all the moments are added up to get the Total Moment. The Total Moment is then divided by the Gross Weight to get the total arm, which is the Center of Gravity.
The calculated values are then entered into the Center of Gravity Limits table that can be found in the Limitations section of the POH.
If all values lie within the envelopes the weight and balance for the helicopter is within the approved limits.

Using the graph method is similar to the mathematical method except that the step of dividing the Total Moment by the Gross Weight is omitted and instead the Total Moment and the Gross Weight are entered into the Allowable Moment vs. Gross Weight Envelope table, which can be found in the Weigh and Balance section of the POH.
Another major difference is that when using the graph method for the R44 only the longitudinal CG position is used, as the manufacturer states that due to close positioning of all items to the centerline it is usually not necessary to determine the lateral CG position.
In case of unusual loading or equipment installation the position of the lateral CG should be checked as well.

Examples for both tables can be found on the next pages.
For the calculation of weight and balance different units must be
used and converted between each other.

Conversion tables for all necessary units can be found in the
conversion section of this book.

Loading &
Performance

Example A - The mathematical Method

Using the mathematical method to calculate weight and balance for a Raven II helicopter with the following details:

Model: R44 Raven II

Empty Weight: 1521.6 lbs.
Longitudinal Arm: 106.0 inches
Lateral Arm: -0,2 inches

Pilot Weight: 185 lbs.
Co-Pilot Weight: 170 lbs.
Right rear
Passenger weight: 165 lbs.
Left rear
Passenger weight: 170 lbs.

Fuel in main tank: 20 USG (120 lbs.)
Fuel in aux tank: 15 USG (90 lbs.)

All doors installed.
Dual Controls installed.

All weights have been added up in the following table to get the Gross Weight, which is 2421.6 lbs..

The Max Gross Weight for the Raven II is 2500 lbs., therefore the weight is within the limitations.
Each weight item is then multiplied with its relating longitudinal and lateral arm to get the resulting longitudinal and lateral moments.

All the longitudinal moments are then added up to get the total longitudinal Moment (227394.6) and all lateral moments are added to get the total lateral Moment (-326.3).

The Total Moment is then divided by the Gross Weight to get the total arm, which is the Center of Gravity.

The calculated values are now entered into the Center of Gravity Limits Table to verify that all values lie within the envelopes and the weight and balance for the helicopter is within the approved limits. Connecting the marks in the Center of Gravity Limits Table shows the values for the time between Take-Off and Empty Fuel.

In this example all values lie within the envelope and therefore the Helicopter is within it's CG Limits.

Item	Weight (lbs.)	Long. Arm	Long. Moment	Lateral Arm	Lateral Moment
Basic Empty Weight	1521.6	106.0	161289.6	-0.2	-304.3
Front right seat	185.0	49.5	9157.5	12.2	2257.0
Baggage	0.0	44.0	0.0	11.5	0.0
Front Left seat	170.0	49.5	8415.0	-10.4	-1768.0
Baggage	0.0	44.0	0.0	-11.5	0.0
Rear right seat	165.0	79.5	13117.5	12.2	2013.0
Baggage	0.0	79.5	0.0	12.2	0.0
Rear left seat	170.0	79.5	13515.0	-12.2	-2074.0
Baggage	0.0	79.5	0.0	-12.2	0.0
Front right door	0.0	49.4	0.0	24.0	0.0
Front left door	0.0	49.4	0.0	-24.0	0.0
Rear right door	0.0	75.4	0.0	23.0	0.0
Rear left door	0.0	75.4	0.0	-23.0	0.0
Removable cyclic	0.0	35.8	0.0	-8.0	0.0
Removable collective	0.0	47.0	0.0	-21.0	0.0
Removable Pedals	0.0	16.8	0.0	-9.5	0.0
Zero Fuel	2211.6	92.9	205494.6	0.06	123.7
Main Fuel Tank	120.0	106.0	12720.0	-13.5	-1620.0
Aux Fuel Tank	90.0	102.0	9180.0	13.0	1170.0
Take-Off	2421.6	93.9	227394.6	-0.13	-326.3

CENTER OF GRAVITY LIMITS
RAVEN II

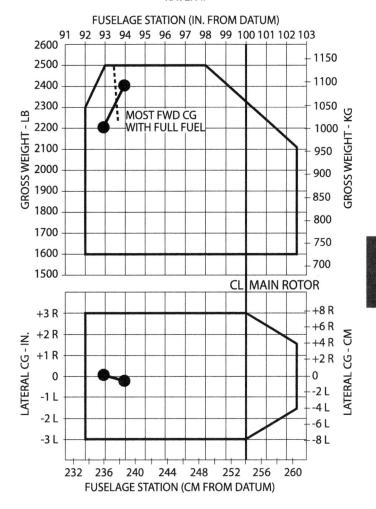

FUSELAGE STATION (IN. FROM DATUM)

MOST FWD CG
WITH FULL FUEL

CL MAIN ROTOR

FUSELAGE STATION (CM FROM DATUM)

Example B - The Graph Method

Using the graph method to calculate weight and balance for a
Raven II helicopter with the following details:

Model:	R44 Raven II
Empty Weight:	1521.6 lbs.
Longitudinal Arm:	106.0 inches
Lateral Arm:	-0,2 inches
Pilot Weight:	185 lbs.
Co-Pilot Weight:	170 lbs.
Right rear	
Passenger weight:	165 lbs.
Left rear	
Passenger weight:	170 lbs.
Fuel in main tank:	20 USG (120 lbs.)
Fuel in aux tank:	15 USG (90 lbs.)

All doors installed.
Dual Controls installed.

All weights have been added up in the following table to get the Gross Weight, which is 2421.6 lbs..

The Max Gross Weight for the Raven II is 2500 lbs., therefore the weight is within the limitations.

Each weight item is then multiplied with its relating longitudinal arm to get the resulting longitudinal moment.
All the longitudinal moments are then added up to get the total longitudinal Moment (227394.6).

The calculated values for Total Moment and Gross Weight are now entered into the Allowable Loaded Moment vs. Gross Weight Envelope Table to verify that all values lie within the envelopes and the weight and balance for the helicopter is within the approved limits.

Important to remember is that the Moment has to be divided by 1000 before using with the table because of its scale.

The calculated values are now entered into the Allowable Loaded Moment vs. Gross Weight Envelope Table to verify that all values lie within the envelopes and the weight and balance for the helicopter is within the approved limits.
Connecting the marks in the Allowable Loaded Moment vs. Gross Weight Envelope Table shows the values for the time between Take-Off and Empty Fuel.

In this example all values lie within the envelope and therefore the Helicopter is within it's CG Limits.

Loading & Performance

Item	Weight (lbs.)	Long. Arm	Long. Moment
Basic Empty Weight	1521.6	106.0	161289.6
Front right seat	185.0	49.5	9157.5
Baggage	0.0	44.0	0.0
Front Left seat	170.0	49.5	8415.0
Baggage	0.0	44.0	0.0
Rear right seat	165.0	79.5	13117.5
Baggage	0.0	79.5	0.0
Rear left seat	170.0	79.5	13515.0
Baggage	0.0	79.5	0.0
Front right door	0.0	49.4	0.0
Front left door	0.0	49.4	0.0
Rear right door	0.0	75.4	0.0
Rear left door	0.0	75.4	0.0
Removable cyclic	0.0	35.8	0.0
Removable collective	0.0	47.0	0.0
Removable Pedals	0.0	16.8	0.0
Zero Fuel	2211.6		205494.6
Main Fuel Tank	120.0	106.0	12720.0
Aux Fuel Tank	90.0	102.0	9180.0
Take-Off	2421.6		227394.6
Moment / 1000			227.39

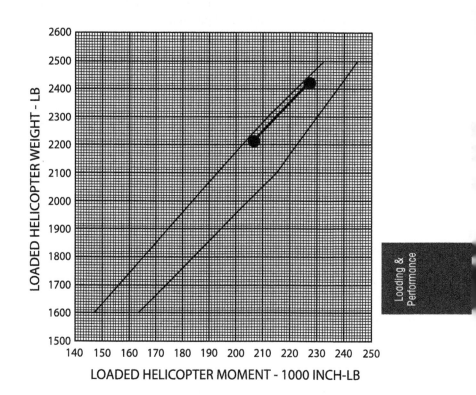

RAVEN II
ALLOWABLE LOADED MOMENT VS. GROSS WEIGHT
ENVELOPE

Performance

The performance expected to be available can be calculated using
the provided performance tables that are provided in the performance
section of the helicopter's rotorcraft flight manual.
Since these tables are based on the results of flights with experienced
test pilots in ideal conditions flight operation should not be performed
all the way to the limits, as allowed by the table, since pilots not as
experienced might not be able to maintain controllability to this point.
Another consideration is that the helicopters that are used for these
tests might be factory-new helicopters, as opposed to helicopters with
more hours that are usually flown in every day life.

General

Hover controllability has been substantiated in 17-knot wind from any
direction up to 9800 feet density altitude.
Refer to the IGE hover performance data for allowable gross weight.

Indicated airspeed (KIAS) shown on the graphs assumes zero
instrument error.

Satisfactory engine cooling has been demonstrated to an outside air
temperature of 38°C (100°F) at sea level or 23°C (41°F) above ISA
at altitude.

Airspeed Calibration Curve

To correct indicated airspeed to calibrated airspeed (assuming no instrument error) the following table is used.

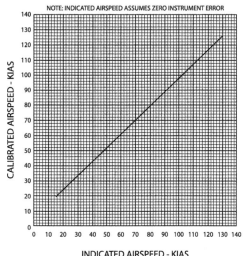

Raven II

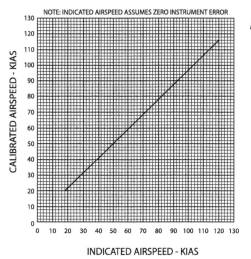

Astro and Raven I

Density Altitude Chart

Density Altitude can be calculated by entering Pressure Altitude and Temperature into the Density Altitude Chart.

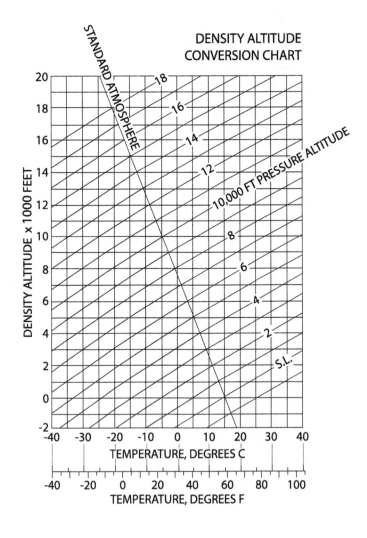

In Ground Effect
at 2 foot skid height

The IGE Hover Ceiling vs. Gross Weight table can be used in multiple ways.

First you can determine the IGE Hover Ceiling for a specifically loaded helicopter in certain conditions.
This is done by starting at the bottom of the table (top for KGs instead of lbs.), selecting the respective gross weight you have determined or planned for the helicopter to have.
From here vertically move up (down if using KGs) until you encounter the respective temperature line, as measured or expected.
Then horizontally move to the left and read the IGE Hover Ceiling as a Pressure Altitude.

Second you can also determine the maximum Gross Weight that still enables the helicopter to hover at a specific altitude. For this calculation choose a Pressure Altitude on the left side of the table. Go horizontally to the left until encountering the respective temperature line, as measured or expected.
From there go down vertically and read the maximum gross weight for this condition (go up if you prefer using KGs instead of lbs.).

IN GROUND EFFECT AT 2 FOOT SKID HEIGHT
FULL THROTTLE
101-102 % RPM
ZERO WIND

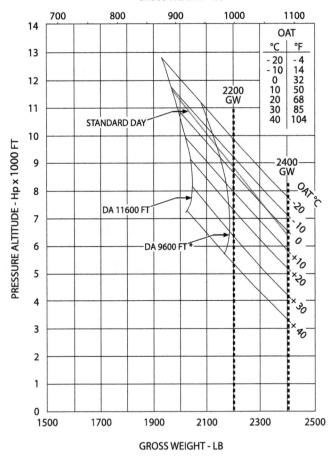

IGE HOVER CEILING VS. GROSS WEIGHT
Astro and Raven I

*Hover contrability with 17 knot wind
substantiated up to 9600 feet density altitude.

IN GROUND EFFECT AT 2 FOOT SKID HEIGHT
FULL THROTTLE
101-102 % RPM
ZERO WIND

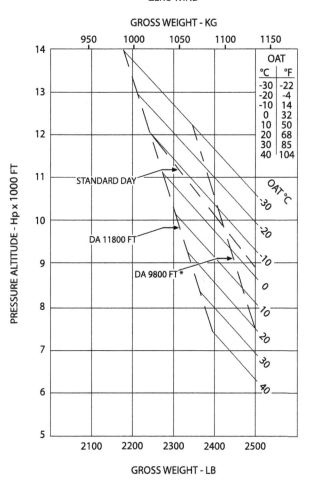

IGE HOVER CEILING VS. GROSS WEIGHT
Raven II

*Hover contrability with 17 knot wind
substantiated up to 9800 feet density altitude.

Out of Ground Effect

The OGE Hover Ceiling vs. Gross Weight table can also be used in multiple ways.

First you can determine the OGE Hover Ceiling for a specifically loaded helicopter in certain conditions.
This is done by starting at the bottom of the table (top for KGs instead of lbs.), selecting the respective gross weight you have determined or planned for the helicopter to have.
From here vertically move up (down if using KGs) until you encounter the respective temperature line, as measured or expected.
Then horizontally move to the left and read the OGE Hover Ceiling as a Pressure Altitude.

Second you can also determine the maximum Gross Weight that still enables the helicopter at a specific altitude.
For this calculation choose a Pressure Altitude on the left side of the table. Go horizontally to the left until encountering the respective temperature line, as measured or expected.
From there go down vertically and read the maximum gross weight for this condition (go up if you prefer using KGs instead of lbs.).

OUT OF GROUND EFFECT
TAKEOFF POWER OR FULL THROTTLE
101-102 % RPM
ZERO WIND

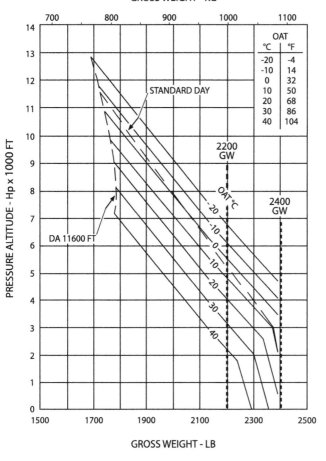

OGE HOVER CEILING VS. GROSS WEIGHT
Astro and Raven I

OUT OF GROUND EFFECT
TAKEOFF POWER OR FULL THROTTLE
101-102 % RPM
ZERO WIND

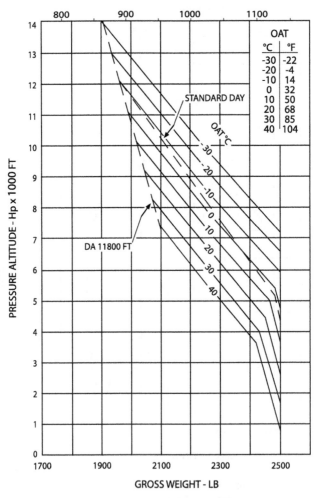

OGE HOVER CEILING VS. GROSS WEIGHT
Raven II

Height-Velocity Diagram

The Height-Velocity Diagram shows safe and unsafe areas resulting of combinations between heights above ground and airspeed settings relating to the ability of performing a successful autorotation to the ground.
The shaded areas are unsafe operation areas and are generally referred to as dead man's curve by pilots as operation in it is likely going to be fatal in case of an engine failure.

There are two main parts of the curve, one resulting of high-speed, low altitude settings and the other one resulting of high altitude but low airspeed settings.
A single-shaded area on the left side is used when operating at high-density altitudes and may be disregarded when operating at sea level.

For example at sea level the minimum height above ground to perform a safe autorotation from 0 KIAS is around 400 feet AGL, while at higher DA's the minimum altitude from 0 KIAS is not below 600 feet AGL.

There is a dashed line, which represents the recommended take-off profile to avoid the shaded areas during takeoff.

Although this table is based on test flights with helicopters loaded close to the maximum gross weight and a pilot in a lighter loaded helicopter may be able to successfully perform an autorotation from inside the shaded areas it is not recommended to fly in these areas.
The shaded areas should be avoided at all times.

Loading & Performance

DEMONSTRATED CONDITIONS:
SMOOTH HARD SURFACE
WIND CALM

AVOID OPERATION IN SHADED AREAS

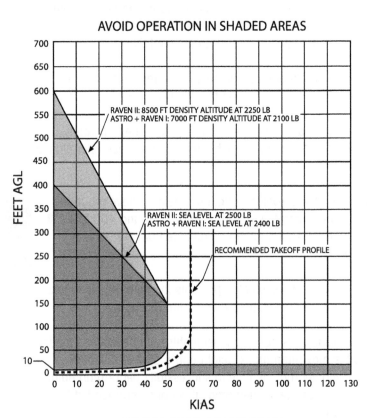

HEIGHT - VELOCITY DIAGRAM

Autorotational Performance

There are certain set values for Autorotational performance described in the Flight Manual.

The maximum glide distance configuration leads to a glide ratio of about 4.7:1, corresponding one nautical mile per 1300 feet AGL. The setup consists of an airspeed of about 90 KIAS and a rotor RPM of about 90%.

The setup for the minimum rate of descent configuration is an airspeed of about 55 KIAS and a rotor RPM setting of about 90%. This leads to a minimum descent of approximately 1350 feet per minute, which corresponds to a glide ratio of about 4:1 or one nautical mile per 1500 feet AGL.

If autorotating below 500 feet AGL rotor RPM should be increased to a minimum of 97%.

Chapter 7

Systems Description

Chapter 7
Systems Description

The Airframe

The R44 airframe is primarily a metal construction.
The primary fuselage structure is welded steel tubing and riveted aluminum sheet. The tailcone is an aluminum semi-monocoque structure where the skin carries most loads.
Attached to it's end are a vertical and a horizontal stabilizer for increased cruise flight stability.
The secondary cabin structure, engine cooling shrouds, the doors and various other ducts and fairings are constructed of fiberglass and thermoplastics.
A stainless steel firewall is located above the engine and a second one forward of the engine separating the engine bay from the cabin.

There are four cowl doors on the right hand side of the helicopter providing access to the main rotor gearbox, the drive system and the engine. Another cowl door on the left hand side provides access to the engine oil filler and dipstick.
For maintenance purposes there are removable panels between the seat cushions and seat backs allowing additional access to controls and other components.
More removable parts are located on both sides and aft of the engine, as well as beneath the cabin.

The instrument console hinges up to allow access to the instruments, wiring and nose-mounted batteries.

Systems
Description

The Landing Gear

Spring and yield skid type landing gear is used on all R44 models.
This system enables the landing gear to absorb most hard landings
elastically, preventing damage to the helicopter's main fuselage.
Extremely hard landings however will cause the struts to hinge up
and outward and the center crosstube to yield in order to absorb the
impact.
The center crosstube will remain in this position thereafter.
While slight crosstube yielding is still acceptable, a yielding that allows the tail skid to be within 30 inches of the ground when the ship
is sitting empty, on level pavement, requires a crosstube replacement.

To protect the skids from excessive
use shoes made of hardened steel are
mounted on the bottom of the skids.
These shoes should be inspected
regularly, even more so if the helicopter
is used for autorotations with ground
contact and/or running landings.
Whenever the thinnest point is less
than 1/16 of an inch (.06 in.) the
shoes should be replaced.

The skid-type landing gear.

The Rotor Systems

The main rotor system used in the R44 consists of two main rotor blades and one forged-aluminum hub. The blades are mounted to the hub by individual coning hinges, while the hub is mounted to the shaft with a teeter hinge.
All three hinges use self-lubricated bearings.

The original main rotor blades were constructed of two stainless steel skins with a honeycomb core bonded to a "D" type of stainless steel spar, and an aluminum forged root fitting, which was enclosed in an oil filled and sealed housing at the blade root.
The thick leading edge spar resisted corrosion and erosion created by dust, sand or rain. The stainless steel skins were bonded to the spar approximately one inch aft of the leading edge. Since this bond may be damaged if the bond line is exposed the blades had to be refinished if the paint eroded and bare metal was visible at the bond line.
A steel forged spindle located in the blade root fitting was used for pitch changing of the blades and also was the attachment for the blade to the main rotor hub. It also contained an integral tusk that contacted the rotor-shaft-mounted droop stops when the blades were at rest or turning at a low RPM to prevent them from teetering.
As standard Raven II series has been equipped with slightly different rotor blades compared to the other models, featuring an increased lifting area on the main rotor blades and aerodynamic tip caps on main and tail rotor blades that reduce the 500-foot flyover noise level by nearly one decibel.

In 2010 new main rotor blades were introduced that are now manufactured using aluminum instead of stainless steel for better dent protection and corrosion resistance at the blade tip. The manufacturing process, as described above, is still valid, except for the replacement of stainless steel with aluminum.

Systems
Description

The new blades are now used on all Raven II's, Raven I's and all Astros that are equipped with hydraulic controls.

The reason Astro models without hydraulic controls are not certified to use the new blades is that Robinson Helicopter Company did not test the new blades with non-hydraulic Astros.

Note: RHC does not own a non-hydraulics R44 anymore.

The Company will continue to manufacture the stainless steel blades for use with non-hydraulic Astros for few more years (as of 2011).

After this time non-hydraulic Astros will have to be upgraded to hydraulic controls.

The tail rotor system consists of a teetering hub with a fixed coning angle, to diminish bending stress on the components and two blades.

The hub contains two self-lubricating bearings, which allow the rotor to teeter.

The blades are constructed of a wrap-around aluminum skin bonded to honeycomb and a forged aluminum root fitting that contains two self-lubricating bearings (non-replaceable) that enable the blades to change pitch.

To prevent the tail rotor from teetering a shaft-mounted polyurethane bumper is used, which contacts the machined hub surface when the teeter limit is reached.

Originally developed for the Raven II new tail rotor blades with round tips are now used with all R44 models.

An interesting side note for pilots who also know the Robinson R22 helicopter is that the R44's tail rotor direction of rotation is clockwise opposed to the counterclockwise rotating R22 tail rotor.

This results in increased tail rotor performance, because the tail rotor's advancing blade therefore is closer to the main rotor downwash and is more efficient. The reason the R22 does not have that advantage mainly is to save weight. The gearbox would have to be heavier, which is ok for the bigger R44 but was not favored in the R22.

The main rotor.

The tail rotor.

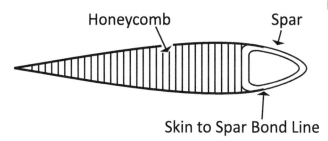

Honeycomb   Spar

Skin to Spar Bond Line

A cross-section of a main rotor blade.

Main Rotor	Raven II	Other models
Articulation	Free to teeter and cone Rigid inplane	Free to teeter and cone Rigid inplane
Number of Blades	2	2
Diameter	33 feet	33 feet
Blade Chord	10.0 inches inboard 10.6 inches outboard	10.0 inches constant
Blade Twist	-6 Degrees	-6 Degrees
Tip Speed @ 102 % RPM	705 FPS	705 FPS

Tail Rotor	Raven II	Other models
Articulation	Free to teeter Rigid inplane	Free to teeter Rigid inplane
Number of Blades	2	2
Diameter	4 feet 10 inches	4 feet 10 inches
Blade Chord	5.1 inches constant	5.1 inches constant
Blade Twist	0	0
Precone Angle	1 Degree	1 Degree
Tip Speed @ 102 % RPM	614 FPS	614 FPS

The Drive System

The engine's power is transmitted to the rotor system through a vee-belt sheave that is mounted onto the engine output shaft using 4 double vee-belts to transmit the engine output to the upper sheave.

The belts are loose before engine start to allow the engine to start without load. After the engine is started the clutch actuator is used to tension the belts to allow power transmission.
A sprag-type overrunning clutch (commonly also referred to as sprag clutch) is contained in the upper sheave's hub and transmits the engine's power forward to the main rotor and aft to the tail rotor via a driveline shaft in its center.
The purpose of the overrunning clutch is to provide both the main and tail rotor with the possibility to continue rotating in the case of an engine failure.
The driveline shaft features flexible couplings at the main gearbox input and at each end of the tail rotor driveshaft to allow for minimal misalignment of the drivetrain.
The Upper sheave reduces the engine's output speed by a 0.778:1 ratio.

The main gearbox, driving the main rotor, contains a splash lubricated single-stage spiral-bevel gear set and is placed above the horizontal firewall, supported by four rubber mounts.
Due to the gear set the main gearbox reduces the drivetrains speed by an 11:57 ratio.
Cooling ducts, which are provided with air by the engine driven cooling fan, provide cooling of the gearbox.

While the long tail rotor driveshaft has no support bearings, it does have a lightly loaded damper bearing.
The tail gearbox, driving the tail rotor, features a splash lubricated single-stage spiral-bevel gear set, just as the main

gearbox does. Different from the main gearbox though, the tail gearbox gear increases the speed of the driveline by a 31:27 ratio.

While the tail rotor input and output shafts are made of stainless steel
to prevent corrosion all other shafts of the drive system are made of
alloy steel.

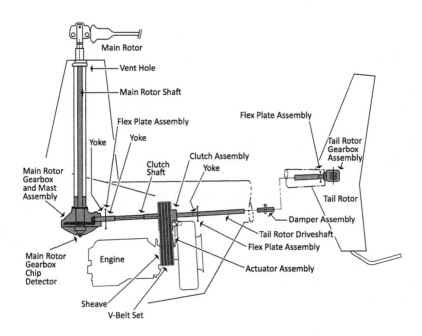

The Powerplant

There are two types of engines used throughout the R44 series:
The Lycoming O-540-F1B5 in the Astro and all Raven I models
and the Lycoming IO-540-AE1A5 in all Raven II models. Both
engines are derated to ensure better engine and transmission
lives.

A major difference between the two engines is that the O-540 is
carbureted, whereas the IO-540 is fuel-injected.

Therefore carb-icing does not exist in Raven II models.

In all models the engine is mounted below the horizontal firewall
and is attached to two forward engine mounts, which are attached to the steel-tube framing. They can be located above and
a little left of the oil dipstick.

An aft support assembly that is attached to the main horizontal
bar of the main welded steel-tube framing further supports the
engine.

*The engine, cased inside the
engine sheet metal.*

Astro and Raven I models

The Lycoming O-540-F1B5 is normally rated at 260 BHP and 2800 RPM but being derated in the R44 it has a maximum continuous rating of 205 BHP at 2718 RPM (equaling 102% on the tachometer) and a 5-minute takeoff rating of 225 BHP, also at 2718 RPM. The engine has 6 cylinders, which are horizontally opposed and staggered, providing each of them with its own crankpin offset from the crankshaft by crank throws.
It is carbureted and uses a wet-sump oil system.
For details on the cooling see the Cooling section on page 7-16.

Induction air is mixed with vaporized fuel as it passes through a venturi in the carburetor. This fuel/air mixture is then delivered to the cylinder intake.
The induction air enters through a screened opening on the right side of the aircraft, next to the rear passenger door, and from there passes to the carburetor-mounted air box assembly via a flexible duct.
Heated induction air is provided by a second flexible duct and drawn by an exhaust-mounted scoop.
The air (cold and/or hot) then flows through the radial-flow air filter and into the carburetor.
The pilot can control the mix of cold and heated air by using the carburetor heat control that is cable-connected to a sliding valve inside the air box.
The problem of carburetor icing is discussed in the mixture in carburetor icing supplement.

Raven II models

The Lycoming IO-540-AE1A5 is normally rated at 260 BHP and 2800 RPM but being derated in the R44 it has a maximum continuous rating of 205 BHP at 2718 RPM (equaling 102% on the tachometer) and a 5-minute takeoff rating of 245 BHP, also at 2718 RPM.
The engine has 6 cylinders, which are horizontally opposed and staggered, providing each of them with its own crankpin offset from the crankshaft by crank throws.
It is fuel-injected and uses a wet-sump oil system.
For details on the cooling see the Cooling section on page 7-16.

Induction air enters through a screened opening on the right side of the aircraft, next to the rear passenger door, and from there flows through the radial-flow air filter within the air box.
From the air box the air then flows along a flexible duct, through the fuel control and into the engine.
Fuel and air are metered at the fuel control unit but are not mixed. The fuel is injected directly into the intake port of the cylinder where it is mixed with the air just before entering the cylinder. This system ensures a more even fuel distribution in the cylinders and better vaporization, which in turn, promotes more efficient use of fuel. Also, the fuel injection system eliminates the problem of carburetor icing and the need for a carburetor heat system.
In case of intake screen or filter blockage a spring-loaded door on top of the air box opens and supplies air from the sheltered engine compartment bypassing the filter.
If this situation should occur some power loss can be expected.

Systems
Description

Cooling (all models)

Air-cooling is used to cool the engine, as well as other parts and
provided via a direct drive, squirrel-cage fan that is attached to the
output shaft of the engine and supplies the cooling air to the muffler,
the main rotor gearbox, the hydraulic reservoir (if installed), the drive
belts and also engine-mounted sheet-metal cooling panels.
These cooling panels then direct cooling air to the drive belts, and
further guide coating air to the cylinders, the external oil cooler (two
on R44 II), the alternator, the magnetos, the fuel flow divider (fuel
injected engines only), and the battery (if the battery is installed in
the engine compartment).

The Induction System

A carbureted system is used in the Astro and Raven I models and a fuel injected system is used in the Raven II models.
The carbureted system mixes the fuel and air in the carburetor before the mixture then enters the intake manifold, whereas the fuel injection system mixes the fuel and air just before the entry into the cylinders.

In the carbureted system, as used in the Astro and Raven I, the induction air enters through a screened opening on the right side of the aircraft, next to the rear passenger door, and from there passes to the carburetor-mounted air box assembly via a flexible duct. Heated induction air is provided by a second flexible duct and drawn by an exhaust-mounted scoop.
The pilot can control the mix of cold and heated air by using the carburetor heat control that is cable-connected to a sliding valve inside the air box. The air then flows through the radial-flow air filter and into the carburetor and through a venturi, a narrow throat in the carburetor.
When the air flows through the Venturi, a low-pressure area is created, which forces the fuel to flow through a main fuel jet located at the throat. The fuel then flows into the airstream where it is mixed with the flowing air. The fuel/air mixture is then drawn through the intake manifold and into the combustion chambers where it is ignited.
The red mixture control is located vertically to the right and front of the cyclic stick.
In flight leaning of the mixture is not allowed in the R44, therefore the mixture control has two positions, full down and full up. In the full down position the mixture is set to full rich, while in the full up position the mixture control is in the idle cut off position and no fuel is delivered to the engine.
A protective cap is supplied and should be set around the mixture knob in flight to prevent accidental pulling of the mixture control,

Systems Description

which would lead to an engine stoppage.

All R44 having a serial number of 202 or higher are equipped with carb-heat assist. The carb-heat control then is correlated to collective settings using a friction clutch.

This means that if the pilot raises collective the carb-heat is reduced, while carb-heat is added, when the pilot lowers the collective.

It is possible for the pilot to override the friction clutch and manually adjust the carb-heat. If carb-heat is not needed it is also possible to lock the control with a latch located at the control knob.

The float-type carburetor acquires its name from a float, which rests on fuel within the float chamber. A needle attached to the float opens and closes an opening at the bottom of the carburetor bowl. This meters the correct amount of fuel into the carburetor, depending upon the position of the float, which is controlled by the level of fuel in the float chamber. When the level of the fuel forces the float to rise, the needle valve closes the fuel opening and shuts off the fuel flow to the carburetor. The needle valve opens again when the engine requires additional fuel. The flow of the fuel/air mixture to the combustion chambers is regulated by the throttle valve, which is controlled by the throttle located on the collective levers.

The main disadvantage of the float carburetor is its icing tendency. Since the float carburetor must discharge fuel at a point of low pressure, the discharge nozzle must be located at the Venturi throat, and the throttle valve must be on the engine side of the discharge nozzle. This means the drop in temperature due to fuel vaporization takes place within the Venturi. As a result, ice readily forms in the Venturi and on the throttle valve. The problem of carburetor icing is discussed more thoroughly in the mixture in carburetor icing supplement.

In a fuel-injected system, as used in the Raven II models, induction air enters through a screened opening on the right side of the aircraft, next to the rear passenger door, and from there flows through the radial-flow air filter within the air box.

From the air box the air then flows along a flexible duct, through the fuel control and into the engine.

Fuel and air are metered at the fuel control unit but are not mixed. The fuel is injected directly into the intake port of the cylinder where it is mixed with the air just before entering the cylinder. This system ensures a more even fuel distribution in the cylinders and better vaporization, which in turn, promotes more efficient use of fuel.

Also, the fuel injection system eliminates the problem of carburetor icing and the need for a carburetor heat system.

The red mixture control is located in an angle on the upper right of the lower instrument panel.

In flight leaning of the mixture is not allowed in the R44, therefore the mixture control has two positions, full down and full up. In the full down position the mixture is set to full rich, while in the full up position the mixture control is in the idle cut off position and no fuel is delivered to the engine. A protective cap is supplied and should be on in flight to prevent accidental pulling of the mixture control, which would lead to an engine stoppage.

In case of intake screen or filter blockage a spring-loaded door on top of the air box opens and supplies air from the sheltered engine compartment bypassing the filter.

If this situation should occur some power loss can be expected.

Systems Description

The Flight Controls

Although the T-bar cyclic appears to be different from the conventional cyclic, the grip moves exactly the same due to a free hinge at the center point. The R44 is equipped with dual controls, that are removable on the left, as standard. All primary controls use push-pull tubes and bellcranks while all bearings used for the control system are either sealed ball bearings or have self-lubricated Teflon liners.

The cyclic pitch control tilts the main rotor disc by changing the pitch angle of the rotor blades in their cycle of rotation. When the main rotor disc is tilted, the horizontal component of lift moves the helicopter in the direction of tilt. Concluding, a forward movement of the cyclic tilts the main rotor disc forward and lets the helicopter move forward, while a lateral movement of the cyclic results in the tilting of the main rotor disc to the according side and the helicopter moving sideward in the desired direction. When flying with forward speed the lateral movement results in the helicopter turning as an airplane would by banking to the desired direction.

A Cockpit view, showing dual controls installed.

The pilot-side cyclic stick of a Raven II model. Note that Astro and Raven I models do not have the alternate starter switch in the cylic grip, that is visible embedded in the right side.

The collective lever changes the pitch evenly on all main rotor blades. When the collective lever is raised the pitch on all blades is increased evenly which results in creating more lift.

The collective levers installed in the R44 are of conventional design with a twist grip throttle, to control the engine RPM, at each end. They are interconnected and actuate the fuel control butterfly valve through a system of bellcranks and push-pull tubes. When the collective is raised, this linkage opens the throttle.

The collective lever with the twist-grip throttle control.

Integrated into the R44's flight controls is a throttle correlator to reduce the pilot workload. A correlator is a mechanical connection between the collective lever and the engine throttle. When the collective lever is raised, power is automatically increased and when lowered, power is decreased. This system maintains RPM close to the desired value, but still requires adjustment of the throttle for fine tuning.

The right collective lever has an engine governor switch on the end, as well as an engine push-to-start button in Raven II's. In addition to the throttle correlator, the electronic governor senses changes in engine RPM and applies corrective throttle inputs through a friction clutch during all powered flight to maintain the engine RPM at 102%. The governor is active above 80% engine RPM only and can be switched on or off using the switch on the pilot's collective lever.

The pilot, if necessary, can easily override the governor by twisting the throttle outboard to increase RPM or twisting it inboard to decrease RPM.

The pilot can roll off the throttle beyond the idle stop prior to a ground contact (run-on) autorotation landing, preventing the throttle from opening when the collective lever is raised.

This is possible because of a detent spring, located in the vertical throttle push-pull tube.

Caution:
When operating at high density altitudes, the governor response rate may be too slow to prevent overspeed during gusts, pull-ups or when lowering the collective.

The antitorque pedals control the pitch, and therefore the thrust, of the tail rotor blades. The main purpose of the tail rotor is to counter-act the torque effect of the main rotor, preventing the helicopter from spinning in the opposite direction of the main rotor.

Besides counteracting torque of the main rotor, by de- or increasing it's thrust. the tail rotor is also used to control the heading of the helicopter while hovering or when making hovering turns. Beginning with the Raven

The pilot side pedals. Note the pins that are used to adjust them for pilot's height.

I model the pedals on the pilot's side are adjustable to pilot's height. A quick release pin is used to easily adjust them.

The collective and the cyclic are both equipped with adjustable frictions.

These frictions are intended for use on the ground and, if applied hold the controls in position, so the pilot can use both hands for start-up and shut-down procedures.

If used in flight caution must be used when applying the frictions to prevent inadvertent blocking of the controls.

The cyclic friction is applied by turning a knob left of the cyclic stick clockwise to increase friction and counterclockwise to decrease friction.

The collective friction lever can be found near the aft end of the pilot's collective. It is actuated aft to increase friction and forward to release it.

Detachable controls on the left seat are standard in the R44.

To remove the left cyclic use the quick-release pin located on the T-bar left of the center. Press the button and pull out the pin at the same time. Then pull the grip out to the left, supporting the stick.

Place the protective cap on the now exposed tube to prevent injury and place the pin back into its original position to prevent loss.

After the left cyclic is removed rotate the right arm clockwise until the stop, depress the stop pin under the cyclic pivot and continue rotating in clockwise manner one turn until the arm is back level. This is done to wind up a balance spring, which is supporting the T-bar for solo usage.

Caution:
Overrotating the cyclic in either wound or
unwound direction will damage the balance spring. To
reinstall the left cyclic follow the reverse procedure.

To remove the left collective lever push the protective boot aft to
reveal the locking pins.
Depress the locking pins and pull out to the front.
To reinstall make sure all placards are facing upward, then use
reverse procedure. Try to gently turn the collective a bit to ensure both
locking pins are fully engaged. Listen for a snapping sound as both
pins are locking.

The left side pedals are removed by depressing the locking pins while
twisting the pedals counterclockwise, then pull the pedal up and out.
To re-install the pedals use the reverse procedure.

Caution:
Above 6000 feet (Raven II) respectively 4000 feet
(Astro and Raven I), throttle correlation and
the governor are less effective.
Therefore, power changes should be slow and smooth.

Caution:
At high power settings above 6000 feet (Raven II)
respectively 4000 feet (Astro and Raven I),
the throttle is frequently wide open and RPM must be con-
trolled with the collective.

The Hydraulic System

Beginning with the Raven I model all R44 helicopters are equipped with hydraulically-boosted main rotor flight controls. Although most of the earlier models have been retrofitted and now feature hydraulic controls as well, there are still some that are not equipped with it. These helicopters then instead of the hydraulic system use an automatic electric trim system, which is described under "Automated Cyclic Trim".

The hydraulic system eliminates feedback forces on the cyclic and collective.
It consists of a pump, three servos, a reservoir and interconnecting lines. To maintain hydraulic pressure in case of an engine failure the pump is mounted on and driven by the main rotor gearbox.
One servo is connected to each of the three push-pull tubes, which support the main rotor swash plate.
The reservoir for the hydraulic fluid is mounted onto the steel tube frame be-
hind the main rotor gearbox. It includes a filter, a pressure relief valve and a pilot-controlled pressure shut-off valve.

The Hydraulic Pump, attached to the Main Rotor Gearbox.

For pre-flight fluid level checks a sight glass is incorporated in the reservoir. This sight glass is accessible via the upper cowl doors on the right side of the helicopter. On top of the reservoir is a vented filler cap.

The pressure shut-off valve is solenoid-actuated and controlled by the hydraulics switch, located on the pilot's cyclic grip.
The switch should remain in the "on" position during all phases of flight and start-up, except for the hydraulic system check or intentional hydraulics-off training. Such hydraulics-off training is highly recommended since controlling the helicopter in a hydraulics-off situation is quite demanding.

The Hydraulic Fluid Reservoir.

One more situation the hydraulics-off switch can be used for is if there is a hydraulics system failure. Since the chance is high not all of the servos fail at the same time but only one or two fail simultaneously the hydraulics would then still work for inputs in certain directions but not for others. Switching off the hydraulics then brings back equal control response. The required pilot input then will be higher due to the lack of hydraulic support.

Note:
- The pressure-shut off valve is a fail-open valve. Electrical power is required to turn off the hydraulics. In case of an electrical failure with the switch being in "off" mode the hydraulics will start working again since there is no more power keeping the valve closed.
- Pulling the HYD circuit breaker will NOT turn off hydraulics but will disable the hydraulic switch.

The Automated Cyclic Trim

Earlier models of the R44 instead of hydraulic controls are equipped with an automated electric trim system.
Although Robinson always knew that eventually they would install hydraulics in the R44 by the time the Astro model was introduced there did not seem to be any hydraulic system on the market that suited the demands Robinson expected the R44 hydraulics to have.
In the end Robinson Helicopters developed their own hydraulic system, which then was fitted in all helicopters beginning with the R44 Raven 1 model.
Since the engineers expected a future upgrade to hydraulics, all Astros with serial numbers from 500 on are fitted with an extra cog in the main gearbox to eventually drive a hydraulic pump.
The standard system for the Astros, however, always remained the automatic electrical trim system.

This trim system senses feedback forces acting on the cyclic and applies compensating trim forces.
The pilot controls for the trim system are located on the cyclic stick.
For fine adjustments of the trim forces the pilot can use the trim control, located above the cyclic.
If using the trim control for adjustments the pilot should wait a few seconds after applying a change to evaluate the result, as there is a lag in response.
The On-Off switch for the trim system is located on the cyclic center post.

Systems
Description

The Ignition System

All R44 engines feature a double ignition system, using two magnetos. The magneto ignition system is the simplest form of ignition system. The engine spins a magnet inside a coil and is therefore creating electromagnetic power.
A Magneto is a small electrical AC generator driven by the crankshaft rotation and creating a very high voltage that is then lead to the spark plugs via high-voltage cables.
The spark plug is located in the cylinder head and is the final destination for the current that is produced from the magneto. By forcing the current to cross a gap a spark is created, which ignites the fuel/air mixture in the cylinder.

Using two magnetos instead of one (most non-aircraft engines only use one) increases safety and reliability and also improves combustion of the fuel/air mixture, which results in a slightly higher power output. If one of the magnetos fails, the other is unaffected. The engine will continue to operate normally, although a slight decrease in engine power can be expected. The same is true if one of the two spark plugs in a cylinder fails.

The operation of the magneto is controlled by the pilot, using the ignition switch.

The switch has five positions:

- OFF
- R (right)
- L (left)
- BOTH
- START

The ignition key switch.

With RIGHT or LEFT selected, only the associated magneto is activated. The system operates on both magnetos with BOTH selected.

A malfunctioning ignition system can be identified during the pretakeoff check by observing the decrease in RPM that occurs when the ignition switch is first moved from BOTH to RIGHT, and then from BOTH to LEFT. A small decrease in engine RPM is normal during this check. The permissible decrease in the R44 is a maximum of 7% in 2 seconds from 75% RPM.

If the engine stops running when switched to one magneto or if the RPM drop exceeds the allowable limit, do not fly the helicopter until the problem is corrected. The cause could be fouled plugs, broken or shorted wires between the magneto and the plugs, or improperly timed firing of the plugs. It should be noted that "no drop" in RPM is not normal, and in that instance, the helicopter should not be flown.

The magnetos are attached to the back of the engine (because of backward engine installation in Robinson helicopters you will find them at the front in R44's) and offset to each side of the engine centerline; therefore they are called the left and the right magneto.

The magneto ignition system is totally independent from the helicopter's electrical system once the engine is running and will continue to work whether battery and alternator are functioning or not.

Systems Description

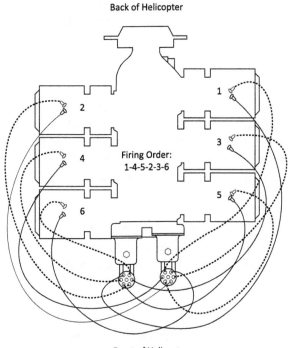

Back of Helicopter

1

2

3

4

Firing Order:
1-4-5-2-3-6

5

6

Front of Helicopter

The Ignition System including the firing order of the cylinders.

The Starter System

In the Astro and Raven I the pilot uses the ignition key switch, by turning into the START position, to start the engine.

The ignition key switch has 5 modes:
- Off - No spark in the engine
- Right - Right magneto on
- Left - Left magneto on
- Both - Both magnetos on (Normal Operation)
- Start - Actuating the Starter

In the Raven II this function has been relocated to starter buttons that have been installed. One is located on the pilot's collective and a second one, mainly intended for air-restart procedures, is located on the pilot's cyclic. The START position of the ignition key switch has been replaced by the PRIME function in the Raven II models.

To start the engine a starter motor is used to initially turn the engine.
The starter motor is battery driven and attached to the engine at the rear end of the engine bay on the pilot's side. It has a cog that engages on to the teeth of a ring gear attached to the engine's crankshaft when the ignition switch is in the START position (Astro, Raven I) respectively one of the START buttons is pressed (Raven II).
Also when the START function is applied by the pilot a starting vibrator is energized and supplies a shower of sparks to starter points in the magneto creating a retarded spark which helps the engine to start.
As soon as the engine works on its own power the pilot releases the ignition key switch (Astro, Raven I) respectively releases the START button (Raven II).
This causes the cog of the starter motor to disengage off the teeth

of the engine's starter ring gear and the starter vibrator to de-energize.

The collective-mounted starter button in a Raven II.

The cyclic-mounted starter button in a Raven II.

Operation of the starter motor is indicated to the pilot by the illumination of the STARTER ON warning light. The light illuminates when the starter motor cog engages to the engine's starter ring gear and stays illuminated until the cog disengages again.

As soon as the ignition key (Astro, Raven I) respectively the starter button (Raven II) is released the light should go out.

If it stays on this indicates a starter motor failure. The starter motor cog then did not disengage and, since the engine is running, is now turned by the engine. This overload situation can cause severe damage and the engine should be shut down by the pilot immediately in this case.

Note:

A starter lockout feature is incorporated to all R44 models as part of the A569 low-rotor-RPM warning unit. Therefore starter activation is only possible when:

1. Clutch switch is disengaged AND (functional) belt tension actuator is fully disengaged.

OR

2. Rotor RPM is above 69%(to allow for in-flight restart).

The Clutch Actuator

As mentioned before the vee-belts, connecting the engine with the upper sheave, and therefore the drive train are loose before the engine starts to enable the engine to start up with no load.
After the engine is started an electric clutch actuator is used to raise the upper sheave and therefore tensions the vee-belts and connects the rotor drive system with the engine.
The clutch actuator is activated by the pilot and activated by engaging the clutch switch.
It uses a sensor to measure belt tension and switches off, as soon as the vee-belts are correctly tensioned.
Whenever the clutch actuator is active, either tensioning, un-tensioning or re-tensioning the belts, this is shown to the pilot by illumination of the clutch warning light, which stays on until the belts are completely tensioned or fully disengaged.

The Clutch actuator, on the bottom of the picture.

In the case of an actuator overload a fuse, located on the test switch panel, prevents the tripping of the circuit breaker, which would result in turning off the clutch light prematurely.

Caution:
Never take off with the clutch
warning light illuminated.

The Fuel System

Older R44 have all-aluminum fuel tanks while newer R44's use flexible bladders in aluminum enclosures in order to reduce the chance of post crash fires.
The bladder version has a slightly reduced capacity as follows:

Fuel Capacity	Total Capacity US gallons (liters)	Usable Capacity US gallons (liters)
Tanks with bladders		
Main Tank	30.5 (115)	29.5 (112)
Aux Tank	17.2 (65)	17.0 (64)
Combined Capacity	47.7 (180)	46.5 (176)
Tanks without bladders		
Main Tank	31.6 (120)	30.6 (116)
Aux Tank	18.5 (70)	18.3 (69)
Combined Capacity	50.1 (190)	48.9 (185)

The fuel tanks are vented through air vents that are located inside the mast fairing.
Both fuel tanks' expansion spaces are interconnected, providing redundancy in the event of the clogging of one of the air vents.
The tanks are interconnected as well, so that the auxiliary fuel tank that is located higher feeds the main tank and will empty while there still remains fuel in the main tank.
From there the fuel flows from the main fuel tank outlet through an on-off valve located on the forward side of the vertical firewall, then through the firewall to a gascolator (the gascolator is a combination of a sediment bowl, a water trap and fine strainer. It separates the fuel from water and sediments). The pilot can control this on-off valve by a knob located between the forward seats.

One electric fuel sender in each tank is connected to its relating fuel quantity gauge in the instrument panel. When the gages read E, the tanks are empty except for a small amount of unusable fuel.

It is recommended to physically check the amount of fuel in the tanks by an optical check or using a dipstick during pre-flight to be able to verify the reading of the fuel gauges.

A float switch in the main fuel tank actuates the LOW FUEL warning light on the upper instrument panel when there are about 3 gallons of usable fuel remaining, this float switch is independent of the fuel senders.

There are drain valves provided for each fuel tank sump, as well as for the gascolator that should be sampled before the first flight of the day and after refueling to ensure no fuel contamination took place and verify the correct fuel grade.

The gascolator drain is located at the lower right side of the vertical firewall, is directly attached to the gascolator and drained by pushing the plastic tube upwards.

For bladder style tanks both, the main and auxiliary tank's fuel drains are located inside the right cowl door below the auxiliary tank.

For aluminum tanks only the auxiliary drain valve is located there, while the drain valve for the main tank is located directly at the exterior forward of the tank on the left side of the aircraft.

When draining the valves located inside the cowl doors the

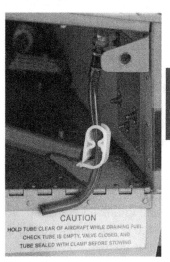

CAUTION
HOLD TUBE CLEAR OF AIRCRAFT WHILE DRAINING FUEL.
CHECK TUBE IS EMPTY, VALVE CLOSED, AND
TUBE SEALED WITH CLAMP BEFORE STOWING

The drain valve located behind the right forward cowl door.

Systems Description

plastic tubes should be extended to be clear of the aircraft, preventing spilling inside the aircraft.

Because of the different types of engines used in the R44 series the operation of the fuel systems past the gascolator are different from each other. The Astro and Raven I's fuel system is gravity fed, while the Raven II's uses a fuel pump system.
In the R44 Astro and R44 Raven I models, a flexible fire-sleeved hose routes the fuel directly from the gascolator to the carburetor fuel inlet, which includes a fine finger screen strainer.

In the Raven II models the system is more advanced.
Here fuel travels from the gascolator to the electric fuel pump, then to the engine-driven fuel pump, then to the fuel control and from there to the flow divider atop the engine via flexible fire-sleeved hoses.
The gascolator fuel filter incorporates a pressure switch, which activates the FUEL FILTER warning light, located at the instrument panel, if the strainer becomes contaminated. If operation is continued with an illuminated fuel filter warning light fuel starvation may occur.
The electric fuel pump is a 30-gph positive-displacement pump and is capable of supplying more fuel than engine demand under all operating conditions.
Excess fuel is allowed to recirculate, providing cooling of fuel system components upstream of the fuel control, by a return hose from the fuel control inlet to the aux fuel tank. In conditions that only require low fuel demand by the engine such as idle, almost all output provided by the electric pump recirculates providing significant cooling and improved hot idle performance.
If the electric pump output pressure falls below 23 psi a pressure switch that is located above the electric fuel pump illuminates the AUX FUEL PUMP warning light in the cockpit. This light also illuminates if the clutch switch is in the disengaged position an the electric fuel pump therefore is deactivated.

A pressure relief valve mounted above the horizontal firewall that maintains return line and thus fuel control inlet pressure at 28 psi controls the recirculation flow.

The engine-driven fuel pump is also capable of supplying more fuel than demanded by the engine under all operating conditions but only has an output pressure of approximately 22 psi. Therefore, if the electric pump is inoperative, the engine will run normally, but the 28 psi relief valve will not open and no fuel recirculation will occur.

The electric auxiliary fuel pump is mainly used to prime the engine for start-up (the ignition switch position PRIME operates the pump for priming prior to engine start) and to provide fuel pump redundancy keeps running in flight as well as long as the engine has oil pressure and the clutch switch is engaged.

Systems Description

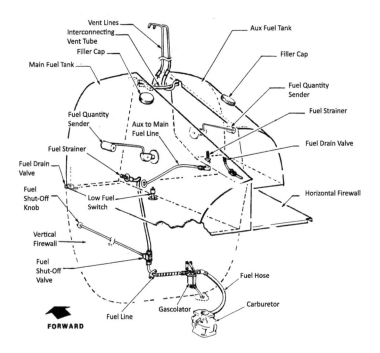

Fuel System Astro + Raven I

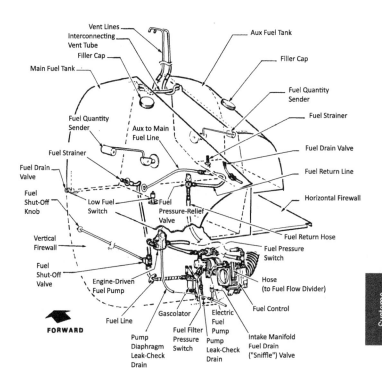

Fuel System Raven II

Systems Description

The Oil System

The engine oil system is used to perform multiple functions that are important for a smooth engine operation which are:

- Lubrication of the engine's moving parts
- Cooling of the engine by reducing friction
- Removing heat from the cylinders
- Providing a seal between the cylinder walls and the pistons
- Carrying away contaminants

The wet sump oil system installed in the R44 models uses an oil pump that draws oil from the oil sump (which is located inside the engine) routes it through a filter screen and, if the oil is hot through the oil cooler(s). From there the oil passes an oil pressure relief valve and is routed into the oil gallery of the crankshaft. The oil flows down into the sump by gravity after it passed through the engine.
The function of the pressure relief valve is to reroute excessive oil directly back to the oil sump in case of too high pressure.

In the Astro and Raven I an oil filter is optional while in the Raven II it is standard.
The oil cooler installed changed over time from earlier Astros that had a rather small oil cooler to a bigger one that was then used as standard in Astros from serial number 0629 on and in all Raven I models. This bigger oil cooler was offered to operators of earlier Astros as a retrofit kit to gain improved cooling margins.
In the Raven II two interconnected oil coolers replace the single one as standard equipment.

For preflight purposes the oil contents can be checked using a dipstick that is accessible when opening the left cowl door.
It is scaled in US quarts and sticks into the oil sump. For appropriate oil contents as required for takeoff and the oil sump capacity check the Limitations section.

When putting the dipstick back into place watch for not fastening it to tight as it might be very hard to take it out again afterwards due to massive temperature changes affecting the material in flight.

In flight the status of the oil system is checked using the oil pressure and the oil temperature gages.
The oil pressure gage provides a direct indication of the oil system operation indicating the pressure of the oil supplied to the engine in pounds per square inch (psi).
There should be an indication of oil pressure during engine start.
The oil temperature gage measures the temperature of the oil. Unlike the oil pressure gage changes in the oil temperature occur more slowly, as can be seen when starting up a cold engine, where it will take several minutes for the gage to see any increase in oil temperature.
The oil temperature gage should be checked periodically during flight as high oil temperature indications may signal a plugged oil line, a low oil quantity, a blocked oil cooler or a defective temperature gage.
A low oil pressure warning light is located on the instrument panel and receives its signal input from the engine oil pressure switch while the oil pressure gage receives its signal from the engine oil pressure sender.

Systems Description

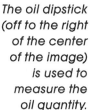

The oil dipstick (off to the right of the center of the image) is used to measure the oil quantity.

The Tachometers

The R44, as all Robinson Helicopters, is not equipped with Tachometers of the conventional concentric type but instead features one electronic dual (engine and rotor) tachometer.
The engine tachometer uses breaker points in the right magneto (left helicopter side) as sensor.

The rotor tachometer uses an electronic setup, based on the Hall effect, to sense the passage of two magnets that are attached to the main rotor gearbox input yoke assembly.

The dual tachometer on the left side.

You can see the sensors as two cylindrical shaped silver pins sticking through the bottom of the rotor brake assembly just next to the yoke as-sembly. Each tachometer circuit has a separate circuit breaker and is com-pletely independent from the other. Either alternator or battery can power them even if the "MASTER BAT" switch is turned off. With alternator, battery and electrical circuits functioning nor-mally tachometer power will be inter-rupted only if "MASTER BAT" and "ALT" switches are turned off and "CLUTCH ENGAGE" switch is disengaged.

The tach sensor located at the main rotor gearbox.

Caution:
The installation of electrical devices can affect the accuracy and reliability of the electronic tachometers, the low RPM warning system and the governor. Therefore, no electrical equipment may be installed in the R44 helicopter unless installation is specifically approved by RHC.

The Electrical System

There are two different electrical systems used throughout the R44 series.

The Astro and Raven I models use a 14-volt electrical system including a 14-volt, 70-ampere capacity alternator (limited to 50 amps continuous), battery relay, alternator control unit and 12-volt battery.

The Raven II, as well as the E.N.G. version models use a 28-volt electrical system including a 28-volt, 70-ampere capacity alternator (limited to 64 amps continuous), battery relay, alternator control unit, and 24-volt battery.

The Police version also uses the 28-volt electrical system but uses a 130- amp capacity alternator that is limited to 85 amps continuous.

The location of the battery varies and is determined by the Factory after having done a weight and balance measurement. The possible locations are in a fiberglass container located on the lower left steel tube frame, in the nose under the upper console, or in the left-front baggage compartment.

In the Police and E.N.G versions the battery is suspended from the tailcone to counteract additional weight in the front of the helicopter due to the installed special equipment.

In normal operation the engine-driven alternator provides electrical power to the electrical system and charges the battery. The alternator produces alternating current (AC), which is then converted into direct current (DC) and supplied to the bus bar that distributes the current to the electrical components on the aircraft.

The main purpose of the battery is to provide energy for starting the engine, as well as enabling the use of systems (e.g. the radio) without having to start the engine.

Another important purpose is to provide an alternate means for

Systems Description

electrical power in case of an alternator failure.

It should be remembered though that the battery's power will be drained rather fast in this case so all non-essential equipment should be switched off and a landing should be performed as soon as practical. The ammeter indicates the current from the battery (a "-" indicates current flowing from the battery and therefore a discharge). The ammeter indicating a discharge and/or the ALT warning light illuminating will indicate an alternator malfunction.

Push-to-reset type circuit breakers are located on the ledge just forward of the left front seat and are marked to indicate their function and amperage.

The MASTER BATTERY switch, located on the console, controls the battery relay, which disconnects the battery from the electrical system.

A small power wire protected by a fuse near the battery bypasses the battery relay. The bypass wire allows the tachometers and the clock to continue to receive battery power with the MASTER BATTERY switch turned off, as long as there is power left in the battery.

The alternator control unit (ACU) senses system voltage at the ammeter shunt via a remote sense wire. The ACU has three functions: it regulates alternator output voltage to maintain a given battery voltage, it warns of low-voltage by illuminating the ALT warning light if voltage decreases below a limit voltage, and it protects against over-voltage by shutting off alternator field if voltage increases above acceptable voltage.

Regulating by the ACU allows the alternator output to more closely follow electrical load demands and reduces voltage fluctuations.

Caution:
Continued flight with a malfunctioning charging system can result in loss of power to the electronic tachometers, producing a hazardous flight condition.

The circuit breakers are located right in front of the left front seat.

The electrical system, as used in the Astro and Raven I.

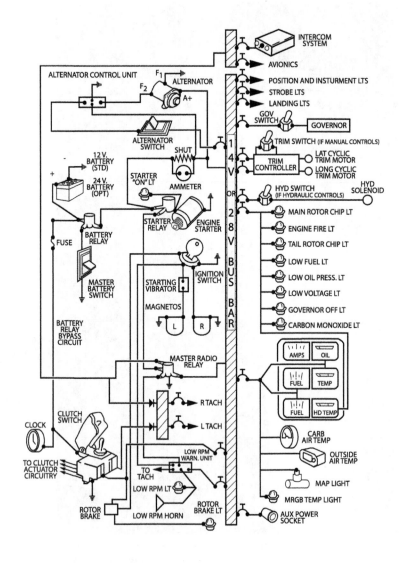

The electrical system, as used in the Raven II.

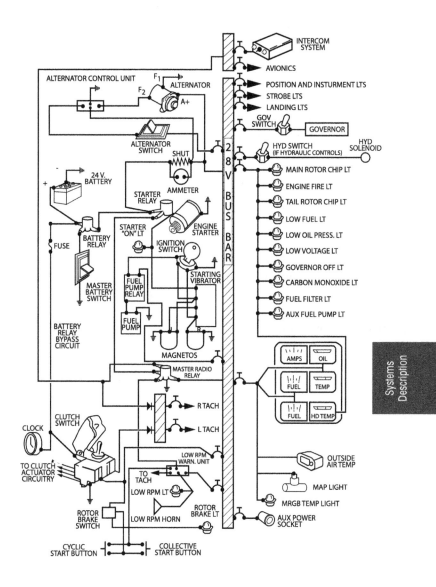

The Lighting System

The R44's lighting system consists of a red anti-collision, twin landing lights, navigation lights, instrument panel lights and a map light.

The anti-collision light is installed on the tailboom and controlled by the STROBE light switch. If there is an optional, additional white anti-collision light installed it is also located on the tailboom.
Then the STROBE light switch function is redefined to operate the white anti-collision light, while the red anti-collision light will be powered whenever the MASTER BATTERY switch is engaged.
Both lights are protected by the STROBE circuit breaker.

Caution:
Turn off the white strobe any time glare is objectionable.

The twin landing lights are installed in the nose of the helicopter's fuselage and are mounted in different vertical angles, one facing more downward than the other, to increase the field that is illuminated and visible to the pilot.
An additional consideration is that the two lights provide for redundancy in case one should fail.
The landing lights are operated via a switch that is located on the cyclic center post. It is not possible to operate them independently, they are either both on or off (except one fails) and have their own LAND LT circuit breaker (used by both lights).

Caution:
The location of the landing light switch should be carefully memorized, so it can be turned on without delay in an emergency.

Note:
The landing lights only operate when the CLUTCH switch is engaged.

The navigation lights are installed below the front doors on both sides of the helicopter (green on the right, red on the left side) and at the end of the tailcone (white).
The pilot activates them by engaging the NAV LTS switch on the instrument console.
Circuit protection is provided by the LTS circuit breaker.

The panel lights illuminate the instruments and are both post and internally mounted.
They function only and always when the NAV LTS switch is engaged. The light intensity can be controlled via a dimmer control located above the NAV LTS switch.
They share the LTS circuit breaker with the navigation lights.

The map light is mounted on the overhead panel and is operated by a switch at its base that allows for OFF and ON operation mode.
Circuit protection is provided by the GAGES circuit breaker.
In some R44 models a LED map light may be installed, which can be recognized by a shorter tube casing.
It then has a three-step switch, allowing for OFF, LOW and HIGH intensity operation mode. In this case an additional LED spotlight is installed at the aft end of the panel, illuminating the rear cabin that is operated by a separate ON, OFF switch.

The two nose-mounted landing lights.

The Intercom System

The R44 features a four-place intercom system, which allows for radio and auxiliary audio input to be mixed with a voice-activated intercom audio.

Pilot and Co-Pilot use cyclic stick mounted trigger-style switches for radio and intercom operation. The triggers have two detents, the first one activating the intercom, while the second activates the transmission function.
In addition there is an intercom switch at each of the 2 rear seats and one in the left forward floor for the passengers, which can also activate the intercom.
The headphone jacks for pilot and co-pilot's headsets are installed at the cabin ceiling. Headset jacks for the rear seat occupants' headsets are located most rearward and centered at the ceiling.

The intercom controls are located on the lower console.
While there are different types of intercoms used in the helicopters they all have the same basic toggles and switches.
These are the ICS VOLUME knob, the VOX SQUELCH knob, the PILOT ISO toggle switch, an indication light and an AUX AUDIO IN jack.

The ICS VOLUME knob is used to control the intercom volume, it does not however control the volume of the radio or the auxiliary audio volume.

The VOX SQUELCH knob can be set to a value between LIVE and MAX and controls the threshold volume that activates the intercom.
When set to LIVE the intercom is always active, while when set to MAX the use of one of the intercom keys is required to activate it.

The PILOT ISO toggle switch is colored red and when PILOT ISO mode is selected disconnects the pilot from the intercom. The pilot then is only connected to the radio, while the passengers and the co-pilot

remain connected on the intercom.

This setting is favorable whenever workload requires the pilot to have a sterile cockpit.

An indication light shows operation of the intercom system by amber illumination and green illumination during radio transmission.

An AUX AUDIO IN jack is located on the rear seat console and accessible for passengers. It can be used to connect personal radios or music players to the intercom system.

Some helicopters, having multiple radios, may feature a slightly different control system that has all of the above features (although the ICS VOLUME and the ICS VOLUME knobs are installed atop each other) and additionally have a MUSIC VOL knob, controlling the volume of devices connected via the AUX AUDIO IN jack.

For the operation of multiple radios there may be a set of toggle switches installed that are used for the monitoring of COM or NAV frequencies. The center position of each switch is the OFF position. Additionally installed then is a transceiver selection knob to choose one of four possible COM transceivers. The fourth COM transceiver selection is titled PA and is used for external amplifiers (e.g. a loud hailer or a passenger address system).

Systems
Description

The front headset jacks are located above the occupant's heads.

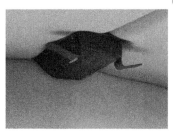

The rear headset jacks are located in a centered console on the ceiling.

The Warning Lights

There are three areas where warning lights are located in all R44's; a row just beneath the upper frame of the instrument panel, another row on top of the lower instrument panel and a single light just above the STROBE switch, which is the BRAKE light, indicating that the rotor brake is applied.

Warning lights located in the upper row are CLUTCH, MR TEMP, MR CHIP, CARBON MONOXIDE (optional in Astro and Raven I, Standard in Raven II), STARTER ON, TR CHIP, LOW FUEL, and LOW RPM.

The ones on the lower row are FUEL FILTER (Raven II only), AUX FUEL PUMP (Raven II only), ALT, ENG FIRE, OIL and GOV OFF.

The CLUTCH light indicates that the clutch actuator is active either tightening or untightening the vee-belts. If on the ground never take off while the clutch light is illuminated.
It is normal for the light to come on momentarily in flight to re-tension

The MR TEMP light indicates an over-temperature in the main gear-box.

The main gearbox chip detector, a magnetic device in its drain plug, activates the MR CHIP warning light. It might happen that metallic particles appear in the gearbox, caused by a failing bearing or gear. These particles then are drawn to the magnetic detector closing an electronic circuit that then illuminates the MR CHIP light.
The TR CHIP light features a corresponding system for the tail gear-box.

A sensor located above the pilot's heater outlet that senses the carbon monoxide levels in the cabin air activates the CARBON MONOXIDE warning light (if installed) when the levels are too high and can be harmful to the occupants.

The STARTER ON illuminates when the starter motor cog engages to the engine's starter ring gear and stays illuminated until the cog disengages again.
If the STARTER ON light stays illuminated after the starter button respectively the ignition key switch is released from the "start" position.

When there are about 3 gallons of useable fuel remaining (correlating to about 10 minutes, using cruise power) a float switch in the main tank activates the LOW FUEL warning light. The warning light should in no case be intentionally used as a fuel level indication.

A rotor RPM of 97% or lower actuates the LOW RPM light. This light is connected to a horn that will be actuated with the light. The low RPM warning system should be checked for being operational before each flight and is included in the checklist.

Unique to the Raven II models are the FUEL FILTER and the AUX FUEL PUMP warning lights. While the AUX FUEL PUMP light indicates low pressure in the auxiliary fuel pump, the FUEL FILTER warning light indicates that the filter got contaminated.
If any one of these two lights illuminates in flight not accompanied by other indication of a problem a landing should be done as soon as practical.
If ,however, there should be other indications like erratic engine operation, or if both lights illuminate, a landing should be done immediately.

The ALT warning light indicates low voltage and a possible failure of the alternator. If the light should come on in flight nonessential electric equipment should be turned off and the ALT switch should be switched off and back on after 1 second in order to reset the overvoltage relay.
If, after this, the light stays on, a landing should be done as soon as practical. Prolonged flight without the alternator functioning can result in the loss of the electronic tachometer, which will lead to a dangerous flight condition.

The ENG FIRE light indicates to the pilot that there is a possible fire in the engine compartment.

Illumination of the OIL warning light indicates loss of engine power or oil pressure.
When illuminated the engine tachometer should be checked for a power loss. The oil pressure gage should be checked as well and, if a loss in oil pressure is confirmed a landing should be done immediately as operation without oil pressure can cause serious damage to the engine and might result in engine failure.

The GOV OFF warning light indicates that the engine RPM throttle governor if not working.
The governor switch should be checked for accidental disengagement.

For more Information about emergency procedures check the related section of this book and the approved rotorcraft flight manual, provided by the Robinson Helicopter Company.

Warning lights located on the upper instrument panel.

Warning lights located on the lower instrument panel.

The warning light test switches, located behind the front right cowl-door.

The main rotor gearbox chip detector sensor is mounted to the bottom of the main rotor gearbox.

The Pitot-Static System

The Pitot-static system consists of:
- the Pitot tube (measuring the dynamic pressure)
- the static port(s) (that measure the ambient/atmospheric pressure)
- the tubing connecting named with the appropriate instruments in the cockpit and these instruments themselves: the altimeter, the vertical speed indicator and the airspeed indicator

In the R44 the Pitot-system features a Pitot tube that is mounted to the front of the main rotor mast fairing and two static sources, which are located aft of the rear door on each side of the helicopter.

The Pitot-tube is mounted to the main rotor mast.

The two interconnected static ports are located behind the rear doors on each side of the fuselage.

The altimeter, the vertical speed indicator and the airspeed indicator use the ambient or static pressure from the static ports.
While the altimeter and the vertical speed indicator solely use the static line though, the airspeed indicator uses both the static and the dynamic pressure, which is provided by the Pitot tube, and measure the difference between those two pressures to indicate the airspeed.
The R44's airspeed indicator shows speed in knots and miles per hour.

Both the Pitot tube and the static ports should be checked before each flight to make sure they are clear and undisturbed.

It is worth mentioning however, that it should never be blown into either the Pitot tube or the static ports, as this could damage the instruments.

In case the instruments seem to show wrong indications water might have settled in the connecting lines. A mechanic can drain them through the forward inspection panel underneath the cabin.

The Heating and Ventilation System and the Carbon Monoxide Detector

Fresh air vents are used for cooling and are located in each door, as well as in the nose of the helicopter.

A push-pull control for the nose vent is located on the face of the lower console, pulling it out opening the vents. The air entering through the nose vent is directed along the inside of the windshield and used not only for cooling purposes but also for defogging.

The vents in the doors are opened by rotating a knob and then pulled open.

When closed the rotating knob locks the vents and prevents opening. It is recommended to open the doors vents fully for hovering but only a few inches in cruise to gain maximum ventilation.

To provide cabin heat a system is used that consists of a heat shroud over the muffler, a control valve on the forward side of the firewall, an outlet grill forward of the pilot's tail rotor pedals, and the interconnecting ducts between the components.

The engine-cooling fan, mounted to the back of the engine, supplies the air.

To control the heating the pilot uses a push-pull heat control that is located on the face of the lower console. Pulling the control up increases the heat, while pushing it all the way down turns off cabin heating. The heat control actuates the control valve at the firewall, which directs the air either into the cabin or out an overboard

The cabin heat control on the left and the push-pull control for the air vent on the right side of the lower instrument panel.

discharge on the underside of the cabin (if heating is turned off). The heating system used in the R44 carries dangers, though, resulting from its working principle.

For one the heating system opens a path from the engine into the cabin, crossing the firewall.
Therefore during start-up and if an engine fire is suspected the heating should always be turned off.

Secondly because of the way the heat is gained in this system, having a heat shroud around the muffler, there is always a chance that carbon monoxide might enter the cabin.
Carbon monoxide (CO) is an odorless, colorless and tasteless gas, highly toxic to humans when encountered in higher quantities.
The gas is existent in the muffler system and therefore in case of a leakage or crack in the muffler can easily enter the cabin via the heat shroud and the heating system.
Because the heat shroud is wrapped around the muffler it is not possible to detect a possible crack during pre-flight inspection, therefore caution should always be used when using the cabin heating.

An in earlier R44 models optional and in later models standard, carbon monoxide detector is used to detect elevated carbon monoxide levels inside the cabin and alert the pilot by the illumination of the CARBON MONOXIDE warning light.
The detector performs a self-test, visible by two flashes, is performed every time power is switched on.
A malfunction of the sensor is indicated by a continuing flash every 4 seconds.
If the light comes on the heater should be switched off and the air vents should be opened.

If the light is accompanied by symptoms of CO poisoning (headache, drowsiness, dizziness) a landing should be performed immediately.

A good habit to prevent carbon monoxide poisoning is to always regulate the cabin temperature by using a combination of cabin heating and ventilation.

The sensor, that is located above the pilot's heater outlet is easily damaged by chemicals and therefore should be covered, when cleaning the cabin interior.

Seats and Seat-belts

The R44 is a four-seat helicopter. The seats each have a baggage compartment underneath that can be accessed by hinging
the seat cushion forward. Maximum compartment capacity is
50 lbs., while the total allowed weight per seat (including the
baggage compartment) is 300 lbs., therefore actual maximum
baggage compartment capacity might be lower, depending on
the occupant's weight.
Items stowed in the baggage compartments should be crushable,
as in case of a hard landing the seats are designed to collapse
and thereby reduce the impact on the occupants.
If the baggage in the compartment is mostly non-crushable (e.g.
Jackets) the seat cannot collapse, as intended and the impact
force will be transferred to the occupant.
While the seats are not adjustable each helicopter comes with
a foam cushion, which can be placed behind the pilot's back
to allow for a more forward seating position and in most cases
the ability to reach the pedals, the cyclic stick even in its most
forward position and to use the controls, located on the center
console.

Each seat is equipped with a combined seat-belt and inertia reel
shoulder strap or, for the front seats only, optional four-point harnesses that provide more stability.
While the harnesses allow for free movement of the occupant
in normal conditions the inertia reel will lock in case of sudden
movement, which would occur in an accident, to restrain the occupant in the seat; very much like in a car.
Newer four-point harnesses are equipped with webbing stops,
located above the inertia reels that limit the retraction of the
harnesses to provide a more comfortable usage. They should be
adjusted to be comfortable, yet not to have too much slack.

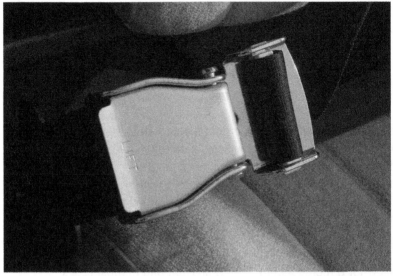

A seat belt buckle.

Note:
The seat belts will only open when the flap is lifted up about 90 degrees, whereas in other aircraft a lifting of only about 30 degrees is sufficient for opening. A passenger briefing should include this fact, as passengers might be lead to believe that their belt is stuck because of this. This could lead to possibly fatal situations following a forced landing.

Doors and Windows

The R44 models are all equipped with 4 doors, so to say one door per seat position.

To secure the doors in the closed position, as well as to lock them for parking a latch system is used.

To close the doors for flight form inside the cabin gently pull the doors shut, then pull the door handle backwards and latch it downward into place. A good habit is to check for successful locking by gently pressing against the door after applying the latch.

To open the doors reverse action is required, first the latch has to be lifted and moved forward and then the door can be pushed open.

All R44 models feature gas springs at each door that open the doors and keep them in the open position eliminating the need to hold them open by hand.

To lock the doors for parking the rear doors are equipped with locking pins, similar to the ones used in cars that are locked by pushing them down from inside the cabin when the door is closed and latched.

If passengers are aboard, sitting in the back, the rear doors should only be latched, not locked, during flight to enable faster evacuation in case of an emergency landing.

The front doors instead feature a key lock on the outside of the door.

Pilots that have flown the R22 might be delighted to notice that in the R44 one key serves for both, the ignition key switch, as well as the door locks.

Systems
Description

Maintenance personnel or the pilot can remove all the doors, although it is recommended to not remove the left doors to prevent objects to be blown out and hitting the tail rotor.
In order to remove a door:

1. Disconnect the door strut by lifting up at the inboard end of the strut, while the door is fully open
2. Remove the cotter rings in the upper and lower hinge pins
3. Then lift the door off

To install the doors use the reverse procedure.

Note:
With any or all doors removed do not exceed 100 KIAS.

The windshield is made of transparent acrylic that is set in a weather-tight silicone sealer and is screwed to the cabin structure. In the R44 they can be either clear or tinted as preferred.
Visibility is mostly un-obscured from all seats due to the relatively small dimensions of the airframe between the window areas in the doors, as well as the windshield.

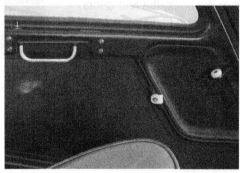

*The left front door,
the latch visible on the left.*

*The hydraulic
door openers.*

The Rotor Brake

The Rotor Brake is mounted on the aft end of the main rotor gear-box and actuated by a cable that is connected to a pull handle located at a panel on the cabin ceiling between the two front seats that also houses the map light.
The handle features a bead chain, allowing the brake to be set for ground handling and parking.

The rotor brake actuator, located in the center above the front seats.

The rotor brake itself is located aft of the main rotor gearbox.

To stop the rotor use the following procedure:

1. After pulling idle cutoff, wait at least 30 seconds.
2. Pull brake handle forward and down using moderate force (10 lb)
3. After the rotor stops, release the handle or, to use as parking brake, hook bead chain into the slot in the bracket.

The Brake must be released before starting the engine.
The starter (Astro, Raven I), respectively the starter buttons (Raven II) are disabled with the rotor brake engaged.

Caution:
Applying the rotor brake without waiting at least
30 seconds after the engine stops or using a force, which stops
the rotor in less than 20 seconds, may permanently damage
the brake shoes.

The Engine Primer System (Optional)

Astro and Raven I models may be equipped with an engine primer system to improve the helicopter's cold starting performance.

The primer pump usually is mounted inside the lower right-side cowl door. In some early helicopters (S/N0357 and earlier) the primer pump is located in the control tunnel; it is then controlled via a knob just forward of the pilot's seat that is connected to the pump by a push-pull cable.

The procedure for engine priming is as follows:

1. Unlock pump handle and pump as required for priming (normally two to three strokes).
2. After priming, push handle full down and lock.

If installed, the primer can be accessed through the lower right cowl door.

The Air Conditioning System (Optional)

The Raven II models offer the option to install a cabin air conditioning system.

The system used is similar to conventional automotive and light aircraft systems and consists of a compressor, accessible through the left engine cowl door, a condenser mounted on the left side of the engine cooling fan scroll, an evaporator and fan assembly mounted to the aft cabin wall, an overhead outlet duct, and interconnecting lines and hoses. The refrigerant used is called R134a.

The compressor is belt-driven from an engine accessory drive cartridge and equipped with an electromagnetic clutch. The compressor clutch is disengaged whenever the system is turned off, allowing the compressor pulley to freewheel.
The evaporator fan draws warm cabin air through the evaporator inlet grill and the evaporator where it is cooled. Cooled air is drawn through the fan and blown through the overhead duct.

The Pilot can control the system using a toggle switch on the overhead duct, which allows selection of off, low, and high fan settings. The compressor is automatically engaged by switching the fan on. A temperature (freeze) switch disengages the compressor when evaporator temperature drops below freezing.
Safety (pressure) switches disengage the compressor if excessive refrigerant leakage occurs or if refrigerant pressure is excessive.
If the engine is near full throttle the compressor is disengaged by a full-throttle switch, to ensure that aircraft performance is not affected. The compressor clutch and fan circuits are protected by the A/C circuit breaker.

The R44 Clippers

The R44 Clippers are specialized versions of the basic models that are float equipped and feature the possibility to land on water.

The most visible difference between the Clipper and the basic model are the floats.

All Clippers can be equipped with fixed floats; the Clipper II can also be equipped with emergency pop-out floats instead.

No matter which kind of floats is used all Clippers feature extended skids and struts enabling the support of the floats.

The emergency pop-out floats are stowed in protective covers along the skid tubes and are identical to the fixed floats, once inflated.

To inflate the pop-out floats, the red inflation lever on the pilot's collective lever must first have its spring-loaded safety in the READY position and then the lever must be squeezed with enough force to shear a special rivet.

For enhanced controllability an additional stabilizer is installed at the base of the lower vertical stabilizer.

The Clippers are further adapted for amphibious operation by additional airframe sealing and corrosion protection.

Fixed-floats clippers also have additional forward position lights installed in the mast fairing, since the floats, when seen from the side, block the main position lights.

A marine radio package is optional.

Note that the V_{NE} is reduced when flying with floats.

The R44 E.N.G. and Police Helicopter

The R44 comes in two specially equipped versions.

The R44 Newscopter is tailored to the needs of electronic news gathering (E.N.G.) and equipped with a nose-mounted gyro-stabilized camera as well as other equipment, including two rear seat equipment panels, that house various audio and video controls and can be fitted to the need of the individual operator by choosing from the wide range of optional equipment.

The R44 Police Helicopter is especially equipped for law enforcement agency usage and features a nose-mounted gyro-stabilized infrared camera with a flat screen LCD monitor and can be equipped with by choosing from a wide range of optional equipment (among others a belly-mounted searchlight, FM transceivers, a video tape recorder, lojack equipment or a microwave transmitter).
To prevent interference with the monitor the left side removable grip has been replaced with a grip on the center post. Therefore the Pilot in Command must occupy the right seat. The Landing gear is extended to provide additional ground clearance for the nose-mounted camera system.

In both configurations the R44 becomes a 3-seat helicopter due to electronic equipment that is installed in and/or on the right rear seat. The left front seat baggage compartment can not be used in the E.N.G. version due to electronic wiring, while the right rear seat baggage compartment can not be used in both versions due to equipment installed here.
Also in the E.N.G. version the weight limit for the left front seat is reduced to 250 lbs. (113 kg).

In order to provide sufficient power for the additional electrical equipment both versions are equipped with 28-volt electrical systems, as well as, in case of the E.N.G. version, an increased-capacity alter-

nator. An additional circuit breaker panel is installed on the ledge just in front of the pilot's seat and contains the circuit breakers for the E.N.G., respectively, the police equipment.

Power to the systems is provided via a NEWS EQUIPMENT MAS-TER SWITCH respectively a POLICE EQUIPMENT MASTER SWITCH that is located on the left of the panel controls.

Due to the installation of the special equipment weight at the nose of the helicopter is increased. To counteract for this, the battery is relocated to a battery box beneath the tailcone in both versions.

Systems Description

Glass Cockpit Retrofit - R44 "Grand"

An Integrated Cockpit Display System (ICDS) called R44 "Grand" can be retrofitted into any model R44. The System has been developed by Sagem and as of May 2011 is certified in the USA, Canada, Brazil, Spain (based on a bilateral agreement with Canada) and Australia. According to the manufacturer one more flight check with a renewed software is required for european certification.

It is designed as a retrofit kit which can replace any installed instrument panel without the need of rewiring the existing R44 harness. The entire Glass Panel/pedestal system is installed by disconnecting the old panel, removing it at the hinge, and replacing it with the updated console. During this process the conventional instruments are replaced by solid-state sensors (Engine Monitoring Module, Air Data Computer ADC, Attitude Heading Reference System AHRS, Tach Gen interface and Sensor interface units), that feed the ICDS system with data.

These replacements can result in advantageous operation of the helicopter due to better reliability, lower operating costs, and safer operation by increased situational awareness. According to the manufacturer the fail rate of the Glass Cockpit instrumentation has been reduced to one in a million compared to one in a thousand with conventional instrumentation. Another advantage of the ICDS system is a weight loss of about 10 lbs. (around 5 kg) compared to a conventional instrument panel.

The system consists of two AMLCD 8-inch displays, that are mounted in portrait mode. One is functioning as a Primary Flight Display (PFD) and the second as a Multifunction Display (MFD).
Integrated are the Primary Flight Instruments, the Engine Instruments and other Multifunction Display system functions such as enhanced vision when linked to a camera, TOPS – Terrain Obstacle Proximity System and others that can be manually arranged on the screens as desired.

The R44 "Grand"
System installed
with the PFD on the
right and the MFD
on the left.

Note the backup
Tachometer and
Altitude Indicator
beneath the displays
for the case of a sys-
tem malfunction.

In this picture the
MFD (on the left) is
set up for a half-
half shared display
showing a moving
map on the bot-
tom half and an
Enhanced Vision
System picture on
the upper half.
This requires the op-
tional installation of
a camera for image
data input.

Chapter 8

Awareness Training

Chapter 8

Awareness Training

Awareness Training

Pilots flying the Robinson 44 in the United States are required by the Federal Aviation Administration to have received Special Awareness training on specified subject matters, which have been identified to be the source of most incidents and accidents related to the R44.

This regulation is called "Special Federal Aviation Regulation No. 73 - Robinson R-22/R-44 Special Training and Experience Requirements" and is commonly referred to as SFAR 73.

The subject areas to be covered are:

• Energy Management
• Low G hazards
• Mast Bumping
• Rotor RPM decay
• Low rotor RPM (blade stall)

Although the SFAR is not necessarily required by Aviation Administrations in other countries it still includes helpful information each pilot should be aware of.

And despite the fact that the SFAR 73 is per definition only applicable to Robinson helicopters the information can be used by pilots flying various kinds of helicopters as most problems are not Robinson specific and can occur in helicopters as big as the Bell 214.

Note that this Awareness Training Information should not be seen as a substitute for instruction by a flight instructor.

Energy Management

Energy management refers to the energy that the helicopter has stored in angular momentum (Rotor RPM), potential energy (Altitude), and kinetic energy (Airspeed). During an autorotation the pilot must use these 3 forms of energy to allow for a safe landing.

In case of an engine failure the first step for the pilot is to lower the collective control to reduce drag on the main rotor blades, which prevents rotor RPM from slowing down.
This causes the helicopter to start descending and thus potential energy is now used to keep the rotor system spinning in exchange for decreasing altitude.

Using the collective, the pilot can move energy between rotor and altitude during the autorotative descent at a set airspeed, to maneuver the helicopter to a landing spot.
Lowering the collective will increase rotor RPM. To spin faster, the rotor system requires energy, so energy is removed from stored altitude and the helicopter's descent rate increases. In reverse, raising the collective takes energy from the rotor system (it slows down) and transfers it to altitude resulting in the helicopter's descent rate decreasing.

However, it is extremely important to not let the rotor RPM get too slow. Allowing this to happen will cause the rotor blades to stall and completely eliminate the pilot's ability to control and use the stored energy.
This catastrophic rotor stall could occur if rotor RPM ever drops below 80 per cent plus 1 percent per 1000 feet of altitude. The helicopter will free-fall and release all its energy at impact—enough energy to destroy the helicopter and its occupants.

Another way the pilot can use the stored energy is by adjusting airspeed in the autorotative glide to reach a landing spot.
Increasing airspeed requires energy, which will be drawn from altitude and rotor RPM, so when increasing the airspeed the helicopter

will descend faster (loss of potential energy) and rotor RPM will drop (loss of angular momentum).
Basically, the pilot is transferring energy from altitude and rotor RPM to airspeed.
Decreasing the airspeed on the other hand transfers energy back to altitude and rotor RPM, causing the sink rate to decrease and the rotor RPM to increase.

It is the skillful manipulation of all this stored energy that will allow the pilot to make a successful power off landing.

As the pilot maneuvers to a landing spot the helicopter gets closer to the ground and is running out of stored altitude energy.
That's OK as the goal is to land.
Maintaining an airspeed of approximately 70 KIAS leaves a healthy amount of energy in airspeed to stop the descent rate.
This is done by flaring at a low altitude, beginning at about 40 feet AGL. During the flare, the rotor system will absorb energy causing RPM to increase and can be controlled by raising the collective.
Care must be taken to not flare too much or add too much collective as this can cause the helicopter to gain altitude. The objective is to bring the helicopter to a momentarily hover-like condition about 8-feet above the surface.
Timed right, all or most of the airspeed energy will be consumed and the helicopter will momentarily be close to the ground with no descent rate and little or no forward speed.
However, it will start descending again and here is where the pilot will raise the collective to provide a burst of lift to cushion the touchdown. Raising the collective uses the energy stored in the rotor system and RPM rapidly slows down. Done right the helicopter will be sitting on the ground with all of its stored energy used.

Low G Hazards And Mast Bumping

For cyclic control, small helicopters depend primarily on tilting the main rotor thrust vector to produce control moments about the aircraft center of gravity (CG), causing the helicopter to roll or pitch in the desired direction. Pushing the cyclic control forward abruptly from either straight-and-level flight or after a climb can put the helicopter into a low G (weightless) flight condition. In forward flight, when a push-over is performed, the angle of attack and thrust of the rotor is reduced, causing a low G or weightless flight condition.

During the low G condition, the lateral cyclic has little, if any, effect because the rotor thrust has been reduced. Also, in a counter-clockwise rotor system (a clockwise system would be the reverse), there is no main rotor thrust component to the left to counteract the tail rotor thrust to the right, and since the tail rotor is above the CG, the tail rotor thrust causes the helicopter to roll rapidly to the right. If, in a 2 bladed helicopter with a teetering hinge, you attempt to correct by applying lateral cyclic input before regaining main rotor thrust however, since the main rotor is unloaded, this only causes the tip path plane to tilt and not the rest of the helicopter and the rotor can exceed its flapping limits and cause structural failure or separation of the rotor shaft due to mast bumping, or it may allow a blade to contact the airframe.

Since a low G condition could have disastrous results, the best way to prevent it from happening is to avoid the conditions where it might occur. This means avoiding turbulence as much as possible.

If you do encounter turbulence, slow your forward airspeed and make small control inputs. If turbulence becomes excessive, consider making a precautionary landing. To help pre- vent turbulence induced inputs, make sure your cyclic arm is properly supported.

One way to accomplish this is to brace your arm against your leg. Even if you are not in turbulent conditions, you should avoid abrupt movement of the cyclic and collective.

If you do find yourself in a low G condition, which can be recognized by a feeling of weightlessness and an uncontrolled roll to the right, you should immediately and smoothly apply aft cyclic. Do not attempt to correct the rolling action with lateral cyclic. By applying aft cyclic, you will load the rotor system, which in turn produces thrust.

Once thrust is restored, left cyclic control becomes effective, and you can roll the helicopter to a level attitude.

Awareness Training

Rotor RPM Decay And
Low Rotor RPM (Blade Stall)

Low rotor RPM occurs any time the rotor tachometer needle is out of
the green arc or normal operating range. If the pilot does nothing to
correct for low rotor RPM it will begin to decay, which can ultimately
lead to a Low rotor RPM blade stall.

Low rotor RPM blade stall is when the rotor RPM is allowed to slow to
a point in which the engine cannot produce enough power to restore
the RPM. If you let rotor RPM decay to the point where all the rotor
blades stall, the result is usually fatal, especially when it occurs at
altitude.

The danger of low rotor RPM and blade stall is greatest in small
helicopters with low blade inertia.

It can occur in a number of ways, such as simply rolling the throttle
the wrong way, pulling more collective pitch than power available, or
when operating at a high density altitude.

When the rotor RPM drops, the blades try to maintain the same
amount of lift by increasing pitch. As the pitch increases, drag increases, which requires more power to keep the blades turning at the
proper RPM. When power is no longer available to maintain RPM,
and therefore lift, the helicopter begins to descend. This changes the
relative wind and further increases the angle of attack. At some point
the blades will stall unless RPM is restored. Catastrophic rotor stall
could occur if rotor RPM ever drops below 80 per cent plus 1 per
cent per 1000 feet of altitude. If all blades stall, it is almost impos-
sible to get smooth air flowing across the blades.

If you are in a low RPM situation, the lifting power of the main rotor
blades can be greatly diminished. As soon as you detect a low RPM
condition, immediately apply additional throttle, if available, while
slightly lowering the collective. This reduces main rotor pitch and
drag.

When in forward flight, gently applying aft cyclic loads up the rotor system and helps increase rotor RPM.

As the helicopter begins to settle, smoothly raise the collective to stop the descent. At hovering altitude you may have to repeat this technique several times to regain normal operating RPM.
This technique is sometimes called "milking the collective."
When operating at altitude, the collective may have to be lowered only once to regain rotor speed.
The amount the collective can be lowered depends on altitude. When hovering near the surface, make sure the helicopter does not contact the ground as the collective is lowered.

Since the tail rotor is geared to the main rotor, low main rotor RPM may prevent the tail rotor from producing enough thrust to maintain directional control. If pedal control is lost and the altitude is low enough that a landing can be accomplished before the turning rate increases dangerously, slowly decrease collective pitch, maintain a level attitude with cyclic control, and land.

Awareness Training

Chapter 9

Carb-Icing Supplement

Chapter 9
Carb-Icing Supplement

Carburetor Icing

A risk when using a carbureted fuel control system, as used in the R44 Astro and Raven I, is the possibility of carburetor icing, which can lead to engine stoppage.

The carburetor's operating principle, based on fuel vaporization and decreasing air pressure in the venturi, causes a sharp drop in temperature within the carburetor, where the temperature in the mixture chamber may drop as much as 21 °C (70 °F) below the temperature of the incoming air.
If the air is moist, the water vapor in the air may condense. When the temperature in the carburetor is at or below freezing, carburetor ice may form on internal surfaces, including the throttle valve.

Because of the sudden cooling that takes place in the carburetor, icing can occur even on warm days with temperatures as high as 38 °C (100 °F) and the humidity as low as 50 percent. However, it is more likely to occur when temperatures are below 21 °C (70 °F) and the relative humidity is above 80 percent.
The likelihood of icing increases as temperature decreases down to 0 °C (32 °F), and as relative humidity increases.
Below freezing, the possibility of carburetor icing decreases with decreasing temperatures.
While special caution should be used regarding carburetor icing in situations where visible moisture (e.g. fog, rain, or operations near water) is present the pilot should never assume carburetor icing not to be a factor when visible moisture is not present.

Although carburetor ice can occur during any phase of flight, it is particularly dangerous when you are using reduced power, such as during a descent. You may not notice it during the descent until you try to add power.
There are multiple indications that can warn the pilot of carburetor icing like a decrease in engine RPM or manifold pressure, the

carburetor air temperature gauge indicating a temperature outside the safe operating range, or engine roughness.
Since changes in RPM or manifold pressure can occur for a number of reasons, it is best to closely check the carburetor air temperature gauge when flying in possible carburetor icing conditions.

Note that the governor may mask the formation of carburetor icing, as it automatically increases throttle and maintains constant manifold pressure and RPM.

The carburetor air temperature gauge is marked with a yellow caution arc reaching from -20 °C to +5 °C (-15 °C to +5 °C in earlier models).
To counteract carburetor icing a carburetor heat system is used, that eliminates the ice by routing air across a heat source(the exhaust manifold), before it enters the carburetor.

Carb heat operation in the R44

During Takeoff:
Due to the fact that helicopters (unlike airplanes) take off using only the power required the risk for carburetor icing increases in this state, even more if the engine and the induction system are still cold.
Therefore full carb-heat should be used during the engine warm-up to preheat the induction system.
Due to the fact that in the R44 both the cold and warm air are filtered there is no risk in using the carburetor heat even in dusty surroundings.
For hover and take off the carb-heat should be applied as necessary to keep the carburetor air temperature gage out of the yellow area.

During Climb or Cruising Flight:
In these conditions the carb-heat should be used as required to keep the carburetor air temperature gage out of the yellow area.

During Descent or Autorotation:
If no visible moisture present - carb-heat should be used as required to keep the carburetor air temperature gage out of the yellow area.
If visible moisture present - full carb-heat should be used.
In helicopters equipped with an older CAT gage where the yellow arc is between -15 °C to +5 °C full carb-heat should be applied when below 18 inches MAP in carburetor icing conditions due to possibly misleading readings of the CAT gage in low power settings.

Carb-Icing
Supplement

Carburetor Heat Assist

Models from and including Serial Number 0202 are equipped with a carburetor heat assist device to help reduce pilot workload.
The carb-heat assist correlates application of carburetor heat with changes in the collective setting.
Heat is added when the collective is lowered (lower power setting), while heat is reduced when the collective is raised (entering high power settings).
The pilot is able to override the assist and reduce or increase carburetor heat as required. If carburetor heat is not required it can be locked off, using a latch located at the bottom of the control knob.
Locking the control knob however is not recommended when the outside air temperature is between 27 °C (80 °F) and -4 °C (25 °F) and the difference between dew point and outside air temperature is less than 11 °C (20 °F).
Carburetor heat should be adjusted as necessary following power changes.

When using the carb heat assist remember that the carb-heat will be reduced as you lift off into a hover and may require readjustment in flight.

Chapter 10

Conversions

Chapter 9
Carb-Icing Supplement

Temperature

° C	° F
100	212
90	194
80	176
70	158
60	140
55	131
50	122
45	113
40	104
35	95
30	86
25	77
20	68
15	59
10	50
5	41
0	32
-5	23
-10	14
-15	5
-20	-4
-25	-13
-30	-22
-35	-31
-40	-40
-45	-49
-50	-58

Conversions

Weight

LBS	KG	KG	LBS
1	0.45	1	2.20
2	0.91	2	4.41
3	1.36	3	6.61
4	1.81	4	8.82
5	2.27	5	11.02
6	2.72	6	13.23
7	3.18	7	15.43
8	3.63	8	17.64
9	4.08	9	19.84
10	4.54	10	22.05
20	9.07	20	44.09
30	13.61	30	66.14
40	18.14	40	88.18
50	22.68	50	110.23
60	27.22	60	132.28
70	31.75	70	154.32
80	36.29	80	176.37
90	40.82	90	198.42
100	45.36	100	220.46
200	90.72	200	440.92
300	136.08	300	661.39
400	181.44	400	881.85
500	226.80	500	1102.31
600	272.15	600	1322.77
700	317.51	700	1543.23
800	362.87	800	1763.70
900	408.23	900	1984.16
1000	453.59	1000	2204.62
2000	907.18	2000	4409.24
3000	1360.77	3000	6613.86
4000	1814.36	4000	8818.48
5000	2267.95	5000	11023.10
6000	2721.54	6000	13227.72
7000	3175.13	7000	15432.34
8000	3628.72	8000	17636.96
9000	4082.31	9000	19841.58
10000	4535.90	10000	22046.20

Conversion Factors

Pound to Kilogram:	x 0.45359
Kilogram to Pound:	x 2.20462

Distance

Meters	Feet		Feet	Meters
1	3.28		1	0.30
2	6.56		2	0.61
3	9.84		3	0.91
4	13.12		4	1.22
5	16.40		5	1.52
6	19.69		6	1.83
7	22.97		7	2.13
8	26.25		8	2.44
9	29.53		9	2.74
10	32.81		10	3.05
20	65.62		20	6.10
30	98.43		30	9.14
40	131.23		40	12.19
50	164.04		50	15.24
60	196.85		60	18.29
70	229.66		70	21.34
80	262.47		80	24.38
90	295.28		90	27.43
100	328.08		100	30.48
200	656.17		200	60.96
300	984.25		300	91.44
400	1312.34		400	121.92
500	1640.42		500	152.40
600	1968.50		600	182.88
700	2296.59		700	213.36
800	2624.67		800	243.84
900	2952.76		900	274.32
1000	3280.84		1000	304.80
2000	6561.68		2000	609.60
3000	9842.52		3000	914.40
4000	13123.36		4000	1219.20
5000	16404.20		5000	1524.00
6000	19685.04		6000	1828.80
7000	22965.88		7000	2133.60
8000	26246.72		8000	2438.40
9000	29527.56		9000	2743.20
10000	32808.40		10000	3048.00

Conversion Factors

Meters to Feet:	x 3.28084
Feet to Meters:	x 0.3048

Conversions

Distance

NM	KM	SM
1	1.85	1.15
2	3.70	2.30
3	5.56	3.45
4	7.41	4.60
5	9.26	5.75
6	11.11	6.90
7	12.96	8.06
8	14.82	9.21
9	16.67	10.36
10	18.52	11.51
20	37.04	23.02
30	55.56	34.52
40	74.08	46.03
50	92.60	57.54
60	111.12	69.05
70	129.64	80.55
80	148.16	92.06
90	166.68	103.57
100	185.20	115.08
200	370.40	230.16
300	555.60	345.23
400	740.80	460.31
500	926.00	575.39
600	1111.20	690.47
700	1296.40	805.55
800	1481.60	920.62
900	1666.80	1035.70

KM	SM	NM
1	0.62	0.54
2	1.24	1.08
3	1.86	1.62
4	2.49	2.16
5	3.11	2.70
6	3.73	3.24
7	4.35	3.78
8	4.97	4.32
9	5.59	4.86
10	6.21	5.40
20	12.43	10.80
30	18.64	16.20
40	24.85	21.60
50	31.07	27.00
60	37.28	32.40
70	43.50	37.80
80	49.71	43.20
90	55.92	48.60
100	62.14	54.00
200	124.27	107.99
300	186.41	161.99
400	248.55	215.98
500	310.69	269.98
600	372.82	323.97
700	434.96	377.97
800	497.10	431.97
900	559.23	485.96

SM	NM	KM
1	0.87	1.61
2	1.74	3.22
3	2.61	4.83
4	3.48	6.44
5	4.34	8.05
6	5.21	9.66
7	6.08	11.27
8	6.95	12.87
9	7.82	14.48
10	8.69	16.09
20	17.38	32.19
30	26.07	48.28
40	34.76	64.37
50	43.45	80.47
60	52.14	96.56
70	60.83	112.65
80	69.52	128.75
90	78.21	144.84
100	86.90	160.93
200	173.80	321.87
300	260.69	482.80
400	347.59	643.74
500	434.49	804.67
600	521.39	965.60
700	608.28	1126.54
800	695.18	1287.47
900	782.08	1448.41

Conversion Factors

Statute Miles to Nautical Miles:	x 0.868976
Statute Miles to Kilometers:	x 1.60934
Kilometers to Statute Miles:	x 0.62137
Kilometers to Nautical Miles:	x 0.539957
Nautical Miles to Statute Miles:	x 1.15078
Nautical Miles to Kilometers:	x 1.852

Volume

Liters	US Gal	IMP Gal
1	0.26	0.22
2	0.53	0.44
3	0.79	0.66
4	1.06	0.88
5	1.32	1.10
6	1.59	1.32
7	1.85	1.54
8	2.11	1.76
9	2.38	1.98
10	2.64	2.20
20	5.28	4.40
30	7.93	6.60
40	10.57	8.80
50	13.21	11.00
60	15.85	13.20
70	18.49	15.40
80	21.13	17.60
90	23.78	19.80
100	26.42	22.00
200	52.84	44.00
300	79.25	65.99
400	105.67	87.99
500	132.09	109.99
600	158.51	131.99
700	184.93	153.98
800	211.34	175.98
900	237.76	197.98

US Gal	IMP Gal	Liters
1	0.83	3.79
2	1.67	7.57
3	2.50	11.36
4	3.33	15.14
5	4.16	18.93
6	5.00	22.71
7	5.83	26.50
8	6.66	30.28
9	7.49	34.07
10	8.33	37.85
20	16.65	75.71
30	24.98	113.56
40	33.31	151.42
50	41.63	189.27
60	49.96	227.12
70	58.29	264.98
80	66.61	302.83
90	74.94	340.69
100	83.27	378.54
200	166.53	757.08
300	249.80	1135.62
400	333.07	1514.16
500	416.34	1892.71
600	499.60	2271.25
700	582.87	2649.79
800	666.14	3028.33
900	749.41	3406.87

IMP Gal	US Gal	Liters
1	1.20	4.55
2	2.40	9.09
3	3.60	13.64
4	4.80	18.18
5	6.00	22.73
6	7.21	27.28
7	8.41	31.82
8	9.61	36.37
9	10.81	40.91
10	12.01	45.46
20	24.02	90.92
30	36.03	136.38
40	48.04	181.84
50	60.05	227.30
60	72.06	272.76
70	84.07	318.22
80	96.08	363.68
90	108.09	409.14
100	120.10	454.60
200	240.19	909.19
300	360.29	1363.79
400	480.38	1818.38
500	600.48	2272.98
600	720.57	2727.58
700	840.67	3182.17
800	960.76	3636.77
900	1080.86	4091.36

Conversion Factors

Imperial Galons to Liters:	x 4.54596
Liters to Imperial Gallons:	x 0.219975
US Gallons to Liters:	x 3.78541
Litres to US Gallons:	x 0.264179
Imperial Gallons to Liters:	x 1.20095
US Gallons to Imperial Gallons:	x 0.832674

Conversions

Index

A

B

C

U

Useful Load 6-5

V

Vee-belt 7-17, 7-19
V_{NE} 2-3, 2-4, 2-14, 7-64

W

Warning And Caution Lights 3-14, 7-50
Window 7-63

Notes

- *check out cfibastian.com*
-
-
-
-
-
-
-
-
-
-
-
-
-
-

CPSIA information can be obtained
at www.ICGtesting.com
Printed in the USA
BVOW11s0208010917
493554BV00009B/112/P